T0135532

Development of Back-Propagation Neural Network for Prediction of Chaotic Data Time Series

(A case study of Indian monsoon rainfall over a smaller geographical region)

Manoj K Kowar   Sanjeev Karmakar   Pulak Guhathakurta

Bibliographic information published by the Deutsche Nationalbibliothek

The Deutsche Nationalbibliothek lists this publication in the Deutsche
Nationalbibliografie; detailed bibliographic data are available
in the Internet at http://dnb.d-nb.de .

©Copyright Logos Verlag Berlin GmbH 2012

All rights reserved.

ISBN 978-3-8325-3304-5

Logos Verlag Berlin GmbH
Comeniushof, Gubener Str. 47,
10243 Berlin
Tel.: +49 (0)30 42 85 10 90
Fax: +49 (0)30 42 85 10 92
INTERNET: http://www.logos-verlag.de

Preface

Long-range monsoon rainfall prediction (forecasting) over a smaller geographical region like "districts" is extremely difficult. Since 1999, India Meteorology Department (IMD), Pune, has been issuing long-range forecasts for three broad homogeneous regions of India, viz. Northwest India, Northeast India, and the Peninsular, only through statistical power regression models. However, it has been found that these models are not satisfactory and it is impractical to study the highly non-linear relationships between long-range monsoon rainfall and its predictors over the districts. Though one can consider the power regression statistical model with ocean or earth surface weather parameters. Basu and Andhariya (Proc. Indian Acad. Sci., Vol. 101, No. 1, pp 27-34, 1992) have found that, the long-range monsoon rainfall data time series of Indian smaller geographical region always representing chaotic in nature with the following features

- No periodic behavior and sensitivity to initial conditions.
- Chaotic motion is difficult or impossible to forecast.
- The motion 'looks' random.
- High Non-linear.

Thus, long-range monsoon rainfall data time series (1951-2004) of Ambikapur region, Chhattisgarh state, India is used for the study. In the Glob, Chhattisgarh state is geographically situated in Latitude $17^0 46'$ N to $24^0 5'$N Longitude $80^0 15'$ E to $84^0 20'$ E. And in the Chhattisgarh, the Ambikapur region of Surguja district (India Meteorological Department station index No. 42693) geographically is located between $21^0 43'$ to $24^0 12'$ N lat., and $81^0 01'$ to $83^0 51'$ E long. The total geographical area (TGA) of this region is 5733 Sq km. In the study four separate, three layers back-propagation neural network (BPN) models have been developed and verified. These models are:

1. BPN model in deterministic forecast.
2. BPN model in parametric forecast.
3. Principal Components BPN model.
4. Hybrid BPN model.

The entire development process of these models as well as their outcomes is provided in this book as in the form of different chapters. Wherein, chapter 1 justifies the applications of ANN technique in non-linear, chaotic motion pattern recognition and prediction. Region of study, methods adopted for collection and pre-processing of data are briefly explained. Chapter 2 deals with the introduction to ANN models. It also discusses in detail the development of ANN models in deterministic, n-parametric, and the training algorithm using Java. ANN models in deterministic forecast, over the Ambikapur region have been developed and verified. The performance of the model have been evaluated and discussed in this chapter 3. In chapter 4, the development of ANN in n-parametric forecast models is explained. An eight parameter ANN model has been developed for long-range monsoon rainfall pattern recognition and prediction for the Ambikapur region. This model is compared with deterministic ANN model as well. Principal component analysis (PCA) is accomplished for the eight parameters intended to reduce input parameter as well as unknown trainable weights. And a separate ANN models have been developed for prediction of monsoon rainfall over the Ambikapur region. The performances of the models are explained by strong relationship between dependent (monsoon rainfall), and independent variables, i.e., PCs. A hybrid ANN model has been developed. Wherein, output of all deterministic, parametric, and PC based parametric model are provided as input to new ANN model. The performance of model is evaluated and also compared with all individual models. It is found that the performance of the model is satisfactory and may be utilize ANN in this manner as well is presented in the chapter 6. And finally, chapter 7 presents the summary of the work. Here, emphasis has been given to conclusions of the work.

Authors (s)

About the Authors

Prof. Manoj K Kowar

 Prof. Kowar did his B.Tech. M.Tech., PhD in Electronics & Telecommunication Engineering & MBA (HRM). Served Birla Institute of Technology, Mesra, Ranchi for 10 years. Currently serving Bhilai Institute of Technology, Durg as Director. Possesses more than 25 years of teaching and research experience. He is also working as a Visiting Professor at IIITM, Gwalior. He has published 157 research papers in various International & National Journals and Conferences. Till date Five research scholars working under Prof. Kowar have been awarded PhD and currently Four research scholars are working under him. His research area includes Secure Communication, Signal Processing, Digital image Processing, Artificial Neural Network, Fuzzy Logic, Bio Medical Instrumentation and Allied Fields. He is a Member and Fellow of many National and International Organizations. He has been awarded the Honorary Membership of International Association of Engineers (UK).

Dr. Sanjeev Karmakar

 Sanjeev Karmakar son of Shri Shital Karmakar & Minu Karmakar, did his B.Sc. MCA., PhD in Computer & Information Technology. Joined Bhilai Institute of Technology (BIT) Durg in the capacity of Associate professor. He has received Young Scientist Award from Chhattisgarh Council of Science Technology-CCOST, Raipur, India in 2007. Other than that he has received five different awards at National level. He has presented & published more than 40 research papers in National and International level Conferences. He has published 20 research papers in National and International Journals. He has published 04 Books for the Institution of Engineers, India. Recently he has published a book by publisher LAP Lambert Acdemic Publishing, Germany entitled "Applications of Neural Network in Long Range Weather Forecasting". He has nominated for publishing his profile in World's who and who's, 2012 Ed., USA. His research area includes, Artificial neural Network, Cryptograph "RSA", Software Engineering, Forecasting. He has organized many National level Conferences. Presently he is guiding 6 PhD, 02 MPill scholars and Co-Guiding 2 PhD scholars. His present responsibilities other then task as associate professor as are, IEEE Transaction Reviewer, Associate Editor International Journal of Artificial Neural Network, Co-Editor of CSVTU

Research Journal, CSVTU Bhilai, Chhattisgarh, India, Secretary IRNet Research Network India Chhattisgarh Chapter, Supervisor and Co-Supervisor of PhD Program in the CSVTU Bhilai and CV Raman University, Bilaspur, Chhattisgarh, India, Expert (Computer Science) Board of Study, Kayan PG Collage, Bhilai, Chhattisgarh India. He is extremely pleased to his Gurudev Prabhat Upadhyay, wife Preeti Karmakar, and loving sons Chinmay & Divyanshu.

Dr. Pulak Guhathakurta

 Guhathakurta did his M.Sc., M. Phil. and Ph. D. in Applied Mathematics from University of Calcutta, India. He was a CSIR Research Fellow at the University of Calcutta, India, before he joined India Meteorological Department (IMD) in 1992. He has gained experience in operational weather forecasting while working at India Meteorological department at Regional Centre, Guwahati as operational weather forecaster. Presently he is working as Scientist and head of the Hydrometeorology Division of India Meteorological Department, Pune. He worked on developing numerical algorithms for solving nonlinear operator equation of special types, facing many computational difficulties and also may not have continuous and invertible derivatives. His field of works also extends on short and long range climate prediction, climate analysis and variability, hydrometeorology, climate changes and its impact on flood risk, flood and drought analysis, application of artificial neural network in climate research etc. One student has been awarded Ph.D under his guidance and presently two students are doing Ph. D under him. He is a recognized teacher and research Guide of the University of Pune. He has more than 30 research papers published in international and national journals, in addition to many research reports and monograph of IMD.

Acknowledgment

We would like to express our gratitude to Bhilai Institute of Technology (BIT) Durg, Chhattisgarh, India Meteorological Department (IMD), Pune, and Chhattisgarh Council of Science & Technology (CCOST), Raipur for providing us a platform, resources to explore and build up this project. We wish to express our thankfulness to Shri I. P Mishra, Honorary Secretary BIT Trust, who offered us the vital infrastructure, and help. The development of this model could not have been done systematically, had the support of Dr. H.P. Das, Deputy Director General, IMD Pune, not been received in time. His help has been especially invaluable in collection of data and literature. The goal was made available due to technical support rendered by Dr. D. S. Pai (Scientist E), IMD Pune, Prof. A. K. Bhattacharya, IIT Khragpur, WB, India, Dr. Das (Scientist E), NBSS&LUP Nagpur, India, Mr. R. N. Bose, Law Officer, Indira Gandhi Agriculture University, Raipur, Chhattisgarh, India.

Contents

Chapter 3 BPN MODEL IN DETERMINISTIC FORECAST 77

Chapters

Chapter 1

JUSTIFICATION OF ANN FOR RAINFALL
FORECASTING

This chapter presents a brief description of Introduction of problem, literature review to be concluded that the artificial neural network (ANN) is sufficiently suitable for prediction of long-range monsoon rainfall (chaos) over a smaller geographical region. This chapter is also presents, region of study, data description and pre-processing ANN model datasets, chaotic behavior of rainfall data as well as other meteorological parameters, methodology of chaos prediction, neural network approach to chaos prediction and finally the conclusions.

1.1   INTRODUCTION

Weather forecasting (especially rainfall) is one of the most important and challenging operational tasks carried out by meteorological services all over the world. It is furthermore a complicated procedure that includes multiple specialized fields of expertise. Researchers in this field have separated weather forecasting methodologies into two main branches in terms of numerical modeling and scientific processing of meteorological data. The most widespread techniques used for rainfall forecasting are the numerical and statistical methods. Even though researches in these fields are being conducted for a long time, successes of these models are rarely visible. The dynamical models are based on the system of nonlinear operator equations governing the atmospheric system. The physics and dynamics of the atmosphere can be better understood by these sets of governing equations. But in the absence of any analogue solution of this system of operator equations, numerical solutions based on approximations and assumptions are the only alternative. Furthermore, the chaotic behaviors of these nonlinear equations, sensitive to initial conditions, make it more difficult to obtain trivial solutions. As a result, there is limited success in forecasting the weather parameters using the numerical model. The accuracy of the models is dependent upon the initial conditions that are inherently incomplete. These systems are not able to produce satisfactory results in local and short-term cases. The performances, however, are poor for long-range prediction of monsoon rainfall even for the larger spatial scale and particularly, for the Indian region. As an alternative, statistical methods in which rainfall time series are treated as stochastic are widely used for long-range predication of rainfall. India Meteorological

Department (IMD) has been using statistical models for predicting monsoon rainfall. Statistical models were successful in those years of normal monsoon rainfall and failed remarkably during the extreme monsoon years like 2002 and 2004. However, it is very difficult to get the same or better skill in predicting district level monsoon rainfall as that of all-India level monsoon rainfall using these statistical models. Two main drawbacks of these statistical models are:

1. Statistical models are not useful to study the highly nonlinear relationships between rainfall and its predictors, even if one considers models like power regression.

2. There is no ultimate end in finding the best predictors. It will never be possible to get different sets of regional and global predictors to explain the variability of the two neighbouring regions having distinguished rainfall features. For example as shown in Fig. 1.1, in two smaller regions of nearby districts of Chhattisgarh, viz., Jagdalpur of Baster district (TGA 14974 and ASR 1313 mm.) and Rajnandgaon district (TGA 8537 and ASR 998 mm.), of Chhattisgarh having contrasting rainfall characteristics, large-scale regional or global predictors have limited role.

The neural network technique is able to get rid of the above two drawbacks. Since 1986, the neural network technique has drawn considerable attention from research workers, as it can handle complex and nonlinear problems better than the conventional statistical techniques. Therefore, proposed work is an attempt to exploit potential of ANN especially in weather parameter pattern recognition as well as prediction

3

by developing deterministic, parametric ANN model and spatial interpolation. Evaluation of these models has been carried out by identifying their degree of relation between dependent and independent variables over the smaller geographical regions like district, subdivision in the context of Chhattisgarh state

Fig. 1.1. Location of Chhattisgarh
Source: IMD Pune, Kalpana-I image/NE Sector of INDIA,
Geographically situated in Latitude $17^0 46'$ N to $24^0 5'$N Longitude $80^0 15'$ E to $84^0 20'$ E

1.2 OBJECTIVES

The work aims at achieving the following objectives:

4

1. To study in detail of ANN technique and its application especially in the field of pattern recognition and prediction of long-range climate variables over very smaller scale geographical regions like district and subdivisions.

2. To search out correlated ocean/earth surface predictors i.e., tele-connected weather parameters those are physically connected with July and monsoon rainfall.

3. To recognize the pattern and predict the monsoon rainfall by using these predictors over the districts and subdivisions.

4. To identify the strength of ANN technique as compared to the statistical technique for prediction.

1.3   REVIEW OF LITERATURE

In much science and engineering practice today, there is an increasing demand for techniques which are capable of predicting of rainfall. These techniques have many applications including rainfall prediction. Mathematically, the general model for prediction of values rainfall 'R' can be expressed as:

$$R = f(x_1, x_2, x_3...x_n, v_1, v_2,....., v_n)$$

Where, $(x_1, x_2, x_3...x_n)$ may be a past recorded rainfall of rainfall prediction parameter i.e., predictor of a perticulaer location and $v_1, v_2, ..., v_n$ are additional variables. In ANN model, dependent variable $x_1, x_2,$

$x_3...x_n$ can be expressed as independent parameter used to input to observe dependent parameter rainfall R, and v_1, v_2,...,v_n can be defined as random weights to be optimized. ANN technique has been successfully applied for the climate prediction over few geographical regions in the globe. This technology is being used to generate the estimator for rainfall at various locations having information from a pattern of surrounding locations. The following section, contributions from the year 1923 to 2012 has been reviewed. It is found that, there are several predictions as well as interpolations by using ANN for solving the above problem has been successfully applied.

After initial work of Walker [1,2] several attempts by Gowariker et al., [3,4], Thapliyal et al., [5,6] have been made for developing better models for long-range forecasts of summer monsoon rainfall in India. Performance of the Gowariker et al., [4]; Rajeevan et al., [7,8], Thapliyal et al., [9] regression models based on different sets of predictors have found to be satisfactory and reasonably accurate during last eleven years. These models are being extensively used by IMD for long-range forecasts of summer monsoon (June-September) rainfall over India as a whole. Recently, IMD has been trying to forecast for Indian sub regions and issuing long-range forecasts for three broad homogeneous regions of India, viz., Northwest India, Northeast India and the Peninsula through the updated three individual power regression models based on different sets of predictors [10]. However, Guhathakurta [11] found these statistical models to be successful in those years of normal monsoon rainfall and failed remarkably during the extreme monsoon years like 2002 and 2004. Also Rajeevan et al., Thapliyal et al., [5, 6, 8-10] have found that the statistical models have many inherent limitations.

Guhathakurta et al., [12] have observed that the correlations between monsoon rainfall and the predictors can never be perfect and there is no ultimate end in finding the best predictors. Parthasarathy et al., Hastenrath et al., [13,14] have found they may suffer epochal changes and there may be cross-correlations between the parameters. Rajeevan et al., Guhathakurta, Krishnamurthy et al., Sahai et al., [9,15-18] have found that attempts to forecast monsoon rainfall as well as climate parameters through statistical technique over smaller areas like a district, or monsoon periods such as a July, monsoon (June-September), have become unsuccessful as correlations fall drastically. Guhathakurta [19,20] has observed that the weather prediction over high-resolution geographical regions is very complicated. However, since 1986, the ANN technique has been drawing considerable attention of research workers, as it can handle the complex non-linearity problems better than the conventional existing statistical techniques.

In the case study of The chaotic time series of Indian monsoon rainfall, Basu and Andharia,1992,have found that the resulting forecast formula uses only the rainfall of past seven years as predictors, making a forecast eight months in advance[21].

After development of recurrent Sigma- Pi neural network rainfall forecasting system, Chow and Cho, 1997, have concluded that the neural network based now casting system is capable of providing a reliable rainfall now casting in Hong Kong [22].

Lee et al., 1998, have found that RBF networks produced good prediction while the linear models poor prediction [23]. Hsieh, W.H. and

B. Tang, 1998, applied various ANN models for prediction and analysis in meteorology data as well as oceanography data and have found ANN technique is extremely useful [24]. In another research Dawson and Wilby have found that, rainfall runoff modeling, the ability of the ANN to cope with missing data and to "learn" from the event currently being forecast in real time makes it an appealing alternative to conventional lumped or semi distributed flood forecasting models [25]. Guhathakurta et al., have found that performance of the hybrid model (model III), has been the best among all three models developed. RMSE of this hybrid model is 4.93%. As this hybrid model is showing good results it is now used by the IMD for experimental long range forecast of summer monsoon rainfall over India as a whole [26].

Ricardo et al., 1999, have used this technology for simulation of daily temperature for climate change over Portugal [27]. Wherein, performances of linear models and non-linear ANN are compared using a set of rigorous validation techniques. Finally, the non-linear ANN model is initialized with general circulation model output to construct scenarios of daily temperature at the present day (1970–79) and for a future decade (2090–99). Charles Jones and Pete Peterson, 1999, [28] have completed a research at the University of California, Santa Barbara, California, for air surface temperature prediction over the city. Guhathakurta, 1999, [29,30] has implemented this technique for short-term prediction of surface ozone at Pune city. In this work multiple regression data analysis using ANN technique has been used. It has been observed that, the parallel model can be developed for all the major cities with different sets of related data but the network architecture will be different.

After comparative study of short term rainfall prediction models for real time flood forecasting, E. Toth et al., have found that the time series analysis technique based on ANN provides significant improvement in the flood forecasting accuracy in comparison to the use of simple rainfall prediction approaches [31]

In 2001 Luk et al., have developed and compared three types of ANNs suitable for rainfall prediction i.e. multilayer feed forward neural network, Elman partial recurrent neural network and time delay neural network [32]. Michaelides et al., have found that ANN is a suitable tool for the study of the medium and long term climatic variability. The ANN models trained were capable of detecting even minor characteristics and differentiating between various classes [33]. After a study of Radial Basis Function Neural Network (RBFNN), Chang et al., 2001, have found that RBFNN is a suitable technique for a rainfall runoff model for three hours ahead floods forecasting [34].

Brath et al., 2002, have presented time series analysis technique for improving the real time flood forecast by a deterministic lumped rainfall runoff model and they have concluded that apart from ANNs with adaptive training , all the time series analysis techniques considered allow significant improvements if flood forecasting accuracy compared with the use of empirical rainfall predictors [35]. Using ANN for daily rainfall runoff modeling, Rajurkar et al., have found that coupling of ANN with a multiple-input single-output model predicted the daily runoff values with high accuracy both in the training and validation periods [36]. For a rainfall runoff relationship Harun and Irwan, 2002, have concluded that the performance of neural network model is better than HEC-HMS and

9

MLR models for modeling the rainfall runoff relationship [37]. Iseri et al., 2002, have developed medium term forecasting of August rainfall in Fukuoka city. In order to identify the sufficient predictors, the partial mutual information was used for the candidate predictors, which are Sea Surface Temperature anomalies (SSTa) in the Pacific Ocean and three climate indices. When data with lead times between one and twelve months were used to forecast August rainfall, it was found that a model with the North Pacific index and selected SSTa as inputs performed reasonably well [38].

Silva and Snell et al., 2003, have applied this approach both in terms of predictive accuracy and model encompassing. This technology is currently being widely applied to climate prediction because of its ability to explain the complex behavior through time series as well as regression data analysis [39, 40]. Richard 2003 has completed simulation of European climate, through this technique. In that study, Neural Network was used for linear regression analysis [41].

In 2004 Maqsood et al., have found that HFM is relatively less accurate and RBFN is relatively more reliable for the weather forecasting problems and in comparison the ensembles of neural networks produced the most accurate forecast [42]. After applying soft computing techniques Pasero and Moniaci, 2004, have found that the system is able to forecast the evolution of the parameters in next three hours giving previous indications about the possibility of rain, ice and fog [43]. Lekkas et al., 2004 have used a multilayer back propagation network and found that BPNN will not always find the correct weight and biases fort the optimum solution, whereas their results supported the hypothesis that

ANNs can produce qualitative forecast. A 7 hour ahead forecast in particular proves to be of fairly high precision, especially when an error prediction technique is introduced to the ANN model [44]. Chang Shu and Donald H. Burn, 2004, have found that artificial neural network ensembles generate improved flood estimates and are less sensitive to the choice of initial parameters when compared with a single artificial neural network [45]. In 2004 Nayak et al., have presented the application of an adaptive neuro-fuzzy inference system (ANFIS) to hydrologic time series modeling, and it was observed that the ANFIS model preserves the potential of the ANN approach fully, and eases the model building process [46]. M.Asce et al., 2004, have applied two ANN-hydrologic forecasting models and they found encouraging results indicating that ANN-hydrologic forecasting models can be considered an alternate and practical tool for stream-flow forecast [47]. In 2004 R.E. Abdel-Aal, studied alternative abductive networks approach, and concluded that the performance is significantly superior to naive forecasts based on persistence and climatology [48]. In coastal areas it is an enormous prediction of tidal level, Tsong-Lin Lee, 2004, has predicted long-term tidal level using back propagation neural network, as compare to conventional harmonic method, he concluded that back-propagation neural network mode also efficiently predicts the long-term tidal levels [49].

To estimate the maximum surface temperature and relative humidity a Feed forward multi-layered artificial neural network model is designed by Chaudhuri, and Chattopadhyay, in 2005, and stated that one hidden-layer neural network is an efficient forecasting tool by which an estimation of maximum surface temperature and maximum relative

humidity can be obtained [50]. A further contribution of Gwo-Fong Lin*
and Lu-Hsien Chen, 2005, in neural network is that, two hidden layers is
developed to forecast typhoon rainfall, and it has been observed that the
forecasting model can produce reasonable forecasts [51]. Using an
innovation in the researches Jon Vandegriff et al., 2005, have studied
Forecasting space weather with ANN and they found that an artificial
neural network can be trained to predict the shock arrival with better
accuracy than existing methods [52]. Ozgur KISI, 2005, has selected
three simple neural network (NN) architectures, i.e. Artificial Neural
Networks, Auto-Regressive Models and sum of square errors, for
comparison of forecasting probabilities and he found that NNs were able
to produce better results than AR models when given the same data
inputs [53]. Exploring the new concept, soft computing models based on
Radial Basis Function Network for 24-h weather forecasting, Maqsood et
al., have concluded that the RBFN produces the most accurate forecasts
compared to the MLP, ERNN and HFM [54].

In 2006, Somvanshi et al., have proved that ANN model can be
used as an appropriate forecasting tool to predict the rainfall, which out
performs the ARIMA (Autoregressive Integrated Moving Average)
model [55]. As in the previous researches we have observed that most of
the researchers have been used Artificial Neural Network for various
annual predictions like rainfall, tide, temperature etc, but in this study
Niravesh Srikalra and Chularat Tanprasert have used Artificial Neural
Network for daily rainfall prediction in Chao Phraya River with Online
Data Collection, and they found that it is possible to predict rainfall on
daily basis with acceptably accuracy using Artificial Neural Network
[56]. A.D. Kumarasiri and D.U.J. Sonnadara, 2006, have applied an

innovative technique for rainfall forecasting using Artificial Neural Networks based on feed-forward back-propagation architecture. Three Neural Network models were developed; a one-day-ahead model for predicting the rainfall occurrence of the next day, which was able to make predictions with a 74.25% accuracy, and two long term forecasting models for monthly and yearly rainfall depth predictions with 58.33% and 76.67% accuracies within a 5% uncertainty level [57]. D. Nagesh et al., 2006, have used Artificial Intelligence techniques for forecasting regional rainfall and they found that this technique shows reasonably good accuracy for monthly and seasonal rainfall forecasting [58]. Guhathakurta, 2006, developed a model for rainfall forecast for the Kerala sub-division based on the area weighted value of all district forecast. The performance was found satisfactory than the statistical technique [59].

As per the utmost necessities of the hydrologists around the globe, Bustami et al., 2007, have studied ANN for precipitation and water level prediction; they found that ANN is an effective tool in forecasting both missing precipitation and water level data [60]. Paras et al., 2007, have introduced a pioneering Feature Based Neural Network Model for Weather Forecasting and the results were very encouraging and it is found that the feature based forecasting model can make predictions with high degree of accuracy [61]. Mohsen Hayati, and Zahra Mohebi, 2007, have used ANN in a new experiment of short term temperature forecasting(STTF) and he found that MLP network has the minimum forecasting error and can be considered as a good method to model the STTF systems [62]. Using time series of draught indices with artificial neural network Morid et al., 2007, have tested number of different ANN

13

models for both Effective Drought Index (EDI) and the Standard Precipitation Index (SPI) with the lead times of 1 to 12 months [63]. As load forecasting is an important prediction aspect for industrial sectors all over the world, Mohsen Hayati, and Yazdan Shirvany, 2007, have put in an approach for short term load forecasting (STLF) using Artificial Neural Network, and they concluded that MLP network has the minimum forecasting error and can be considered as a good method to model the STLF systems [64]. To apply a reliable and robust procedure for monthly reconstruction of precipitation time series, Lucio et al., 2007, have found that Artificial Neural Network (ANN) can be applied to explore the spatiotemporal dependence of meteorological attributes [65]. In another experiment Hartmann et al., 2007, have found that the neural network algorithms are capable of explaining most of the rainfall variability even it can predict the summer rainfall also [66].

Aliev et al., 2008, have proposed, fuzzy recurrent neural network (FRNN) based time series forecasting method for solving forecasting problems, in an experiment and they found that The performance of the proposed method for forecasting fuzzy time series shows its high efficiency and effectiveness for a wide domain of application areas ranging from weather forecasting to planning in economics and business [67]. Chattopadhyay and Chattopadhyay, 2008, have worked out to find out best hidden layer size for three layered neural net in predicting monsoon rainfall in India, and they have found that eleven-hidden-nodes three-layered neural network has more efficacy than asymptotic regression in the present forecasting task [68]. Hung et al., 2008, have developed a new ANN model for forecasting rainfall from 1 to 6 h ahead at 75 rain gauge stations in the study area as forecast point from the data of 3 consecutive years (1997–1999), and they observed that the developed ANN model can be used for real-time rainfall forecasting and flood management [69]. In an comparative study between Artificial Intelligence and Artificial Neural Network for rainfall runoff modeling, Aytek et al.,2008, have found that genetic programming (GP) formulation performs quite well compared to results obtained by ANNs and is quite practical for use. It is concluded from the results that GEP can be proposed as an alternative to ANN models [70]. Chattopadhyay et al.,2008, have studied the complexities in the relationship between rainfall and sea surface temperature (SST) anomalies during the winter monsoon using scatter plot matrices and autocorrelation functions, and they found that the statistical assessment revealed the potential of artificial neural network over exponential regression [71]. However Mar, and Naing, 2008, have tested more over 100 cases by changing the number of input and hidden nodes from 1 to 10 nodes, respectively, and only one output node in an optimum artificial neural network architecture and they concluded that 3 inputs-10 hiddens-1 output architecture model gives the best prediction result for monthly precipitation prediction [72].

Karmakar et al., 2008, 2009 have developed the ANN models for Long-Range Meteorological Parameters Pattern Recognition over the Smaller Scale Geographical Region and the performances of these models in pattern recognition and prediction have been found to be extremely good [73-74].

Hocaoglu et al., 2009, have developed adaptive neuro-fuzzy inference system for missing wind data forecasting [75]. In a Case Study on Jarahi Watershed, Karim Solaimani, 2009, has studied Rainfall-runoff

15

Prediction Based on Artificial Neural Network and he found that Artificial Neural Network method is more appropriate and efficient to predict the river runoff than classical regression model [76]. KOŠCAK et al., 2009, have compared common meteorological forecasting method with ANN and he found the performance of ANN with high accuracy [77]. Karamouz et al., 2009, have experimented to perform long lead rainfall forecasting Using Statistical Downscaling and Artificial Neural Network Modeling; finally they found that the SDSM outperforms the ANN model [78]. In a comparative study between ASTAR and ARIMA methods for rainfall forecasting in Indonesia Otok, and Suhartono, 2009, have concluded that the best model is ASTAR model both in sample and out-sample data [79].

It can be well-known that neural network can applied for most of the prediction aspects, Nekoukar et al., 2010, have used radial basis function neural network for financial time-series forecasting, and the result of their experiment shows the feasibility and effectiveness [80]. Weerasinghe et al., 2010, have tested the performance of neural network, in an experiment, for forecasting daily precipitation using multiple sources, A cluster of ten neighboring weather stations having 30 years of daily precipitation data (1970 – 1999) was used in training and testing the models. Twenty years of daily precipitation data were used to train the networks while ten years of daily precipitation data were used to test the effectiveness of the models. They found that the models were able to predict the occurrence of daily precipitation with an accuracy of $79\pm3\%$ and Fuzzy classification produced a higher accuracy in predicting 'trace' precipitation than other categories [81]. Luenam et al, , 2010 have presented a Neuro-Fuzzy approach for daily rainfall prediction, and their experimental results show that overall classification accuracy of the neuro-fuzzy classifier is satisfactory [82]. Wu et al., 2010 have attempted to seek a relatively optimal data-driven model for Rainfall time series forecasting using Modular Artificial Neural Networks, they found that the normal mode indicate, MANN performs the best among all four models, but the advantage of MANN over ANN is not significant in monthly rainfall series forecasting [83]. To predict the intensity of rainfall using artificial neural network Nastos et al., 2010 have developed prognostic models and they have proved that the results of the developed and applied ANN models showed a fairly reliable forecast of the rain intensity for the next four months [84]. Patil and Ghatol, 2010, have used various ANN topologies such as radial basis functions and multilayer perceptron with Levenberg Marquardt and momentum learning rules for predicting rainfall using local parameters and they found the topologies fit for the same task [85]. Tiron, and Gosav, 2010, have estimated rainfall from BARNOVA WSR-98 D Radar using Artificial Neural Network and the efficiency of ANN network in the estimation of the rain rate on the ground in comparison with that supplied by the weather radar is evaluated [86]. Goyal and Ojha, 2010, have focused their working on a concept of using dimensionless variables as input and output to Artificial Neural Network (ANN), finally they have concluded that ANN model using dimensionless variables were able to provide a better representation of rainfall–runoff process in com-parison with the ANN models using process variables investigated in this study [87]. On the basis of humidity, dew point and pressure in India, Enireddy et al., 2010, have used the back propagation neural network model for predicting the rainfall. In the training they have obtained 99.79% of accuracy and 94.28% in testing. From these results they have concluded that rainfall can predicted in future using the same method [88]. Haghizadeh et al., 2010, have proposed ANN model and Multiple Regression (MR) for prediction of total sediment at basin scale and they found that estimated rate of sediment yield by Artificial neural networks is much better fits with the observed data in comparison to MR model [89]. Subhajini and Raj, 2010, have put in a Computational Analysis of Optical Neural Network Models to Weather Forecasting; in this study they have compared Electronic Neural Network (ENN) model and opto-electronic neural network model. Overall their conclusion was, the training of opto-electronic neural network is fast compared to ANN. The accuracy of optoelectronic neural network is as good as ENN [90]. Durdu Omer Faruk, 2010, has experimented with A hybrid neural network and ARIMA model for water quality time series prediction. He has provided the results that the hybrid model provides much better accuracy over the ARIMA and neural network models for water quality predictions [91]. To identify and forecast the intensity of wind power and wind speed, Soman et al., 2010, have applied artificial neural network (ANN) and hybrid techniques over different time-scales, they found the accuracy in prediction associated with wind power and speed, based on numeric weather prediction (NWP) [92].

Forecasting daily rainfall at Mashhad Synoptic Station, Khalili et al., 2011, have applied Artificial Neural Networks model and they found that the black box model is capable of predicting the rainfall [93]. Pan et al., 2011, have experimented with feed forward neural network to predict Typhoon Rainfall. FNN is applied to estimate the residuals from the linear model to the differences between simulated rainfalls by a typhoon rainfall climatology model (TRCM) and observations and their results

18

were satisfactory [94]. Joshi and Patel, 2011, have put in a review report on Rainfall-Runoff modeling using ANN, in the same study they have reviewed three neural network methods, Feed Forward Back Propagation (FFBP), Radial Basis Function (RBF) and Generalized Regression Neural Network (GRNN) and they have seen that GRNN flow estimation performances were close to those of the FFBP, RBF and MLR [95]. El-Shafie et al., 2011, have developed two rainfall prediction models i.e. Artificial Neural Network model and Multi regression model (MLR). An analysis of two statistical models developed for rainfall forecast on yearly and monthly basis in Alexandria, Egypt shows that an ANN has a better performance than an MLR model [96]. Rainfall forecasting in a mountainous region is a big task in itself Mekanik et al., 2011, have tried to do it using ANN modeling a feed forward Artificial Neural Network (ANN) rainfall model was developed to investigate its potentials in forecasting rainfall. A monthly feed forward multi layer perceptron neural network (ANN) rainfall forecasting model was developed for a station in the west mountainous region of Iran [97]. The temperature has a great effect in forecasting rainfall Amanpreet Kaur, and Harpreet Singh, 2011, have tested Artificial Neural Network in forecasting minimum temperature, they have used multi layer perceptron architecture to model the forecasting system and back propagation algorithm is used to train the network. They found that minimum temperature can be predicted with reasonable accuracy using ANN model [98]. El-Shafie et al., 2011, have proposed an idea of using adaptive neuro-fuzzy inference system based model for rainfall forecasting on monthly basis and they found that the ANFIS model showed higher rainfall forecasting accuracy and low error compared to the ANN model [99]. Tripathy et al., 2011, have experimented with Artificial Neural Network (ANN) and Particle

19

Swarm Optimization (PSO) Technique for weather forecasting and their experimental results indicate that the proposed approach is useful for weather forecasting [100]. As we have observed that many of the scientists have used ANN and various ANN models for forecasting Rainfall, Temperature, Wind and Flood etc., El-Shafie et al., 2011, have compared and studied Dynamic Vs Static neural network models for rainfall forecasting, they have developed AI based forecasting architectures using Multi-Layer Perceptron Neural Networks (MLPNN), Radial Basis Function Neural Networks (RBFNN) and Adaptive Neuron-Fuzzy Inference Systems (ANFIS), finally they concluded that the dynamic neural network namely IDNN could be suitable for modeling the temporal dimension of the rainfall pattern, thus, provides better forecasting accuracy [101]. Geetha and Selvaraj, 2011, have predicted Rainfall in Chennai using back propagation neural network model, by their research the mean monthly rainfall is predicted using ANN model. The model can perform well both in training and independent periods [102]. In various researches Artificial Neural Networks (ANNs) have been extensively used for simulation of rainfall-runoff and other hydrological processes. Reshma et al., 2011, have applied Artificial Neural Network for determination of Distributed Rainfall- Runoff Model Parameters and they found that the ANN technique can be successfully employed for the purpose of estimation of model parameters of distributed rainfall-runoff model [103]. Mohan Raju et al., 2011, have developed Artificial Neural-Network-Based Models for the Simulation of Spring Discharge, as their training and testing results revealed that the models were predicting the weekly spring discharge satisfactorily [104]. Predicting groundwater level is somehow a difficult task now a day because it varies place to place and round the globe, Mayilvaganan, and

Naidu, 2011, have attempted to forecast groundwater level of a watershed using ANN and Fuzzy Logic. A three-layer feed-forward ANN was developed using the sigmoid function and the back propagation algorithm. Now it has been observed that ANNs perform significantly better than Fuzzy Logic [105]. El-shafie et al., 2011, have tried to use neural network and regression technique for rainfall-runoff prediction finally they concluded that the results showed that the feed forward back propagation Neural Network (ANN) can describe the behaviour of rainfall-runoff relation more accurately than the classical regression model [106]. In another research integrated artificial neural network-fuzzy logic-wavelet model is employed to predict Long term rainfall by Afshin et al., 2011. The results of the integrated model showed superior results when compared to the two year forecasts to predict the six-month and annual periods. As a result of the root mean squared error, predicting the two-year and annual periods is 6.22 and 7.11, respectively. However, the predicted six months shows 13.15. [107]. Siou et al., 2011, have experimented with Complexity selection of a neural network model for flood forecasting, these models yield very good results, and the forecasted discharge values at the Lez spring are acceptable up to a 1-day forecasting horizon [108]. Saima et al., 2011, have reviewed on the various forecasting methods and hybrid models, they have put in a contradictory conclusion that there is no such model exists that can forecast accurately in all situations. This is because the distinct nature of the model [109].

At the end of literature survey from 1923 – 2012, In 2012, Sawaitul et al., have presented an approach for classification and prediction of future weather using back propagation algorithm, and

21

discussed different models which were used in the past for weather forecasting, finally the study concludes that the new technology of wireless medium can be used for weather forecasting process. It also concludes that the Back Propagation Algorithm can also be applied on the weather forecasting data. Neural Networks are capable of modeling a weather forecast system [110].

1.3.1. Outcome of the Review

Two main architectures of ANN have been found by the above literature review and those are sufficiently suitable to predict chaotic behavior of monsoon rainfall are discussed in the following subsections-

1.3.1.1. Radial Basis Function Network (RBFN)

Lee et al., 1998, have found that RBFN produced good prediction while the linear models poor prediction [23]. After a study of RBFN, Chang et al., 2001, have found that it is a suitable technique for a rainfall runoff model for three hours ahead floods forecasting [34]. In 2004 Maqsood et al., have found that HFM is relatively less accurate and RBFN is relatively more reliable for the weather forecasting problems and in comparison the ensembles of neural networks produced the most accurate forecast [42]. Exploring the new concept, soft computing models based on RBFN for 24-h weather forecasting, Maqsood et al., have concluded that the RBF neural network produces the most accurate forecasts compared to the MLP, ERNN and HFM [54]. As we know that neural network can applied for most of the prediction aspects, Nekoukar et al., 2010, have used radial RBFN for financial time-series forecasting, and the result of their experiment shows the feasibility and effectiveness [80]. Patil and Ghatol, 2010, have used various ANN topologies such as radial RBF and multilayer perceptron with Levenberg Marquardt and momentum learning rules for predicting rainfall using local parameters and they found the topologies fit for the same task [85]. Joshi and Patel, 2011, have put in a review report on Rainfall-Runoff modeling using ANN, in the same study they have reviewed three neural network methods, Feed Forward Back Propagation (FFBP), RBFN and Generalized Regression Neural Network (GRNN) and they have seen that GRNN flow estimation performances were close to those of the FFBP, RBF and MLR [95]. It is observed that many of the scientists have used ANN and various ANN models for forecasting Rainfall, Temperature, Wind and Flood etc., El-Shafie et al., 2011, have compared and studied Dynamic Vs Static neural network models for rainfall forecasting, they have developed AI based forecasting architectures using Multi-Layer Perceptron Neural Networks (MLPNN), Radial Basis Function Neural Networks (RBFNN) and Adaptive Neuron-Fuzzy Inference Systems (ANFIS), finally they concluded that the dynamic neural network namely IDNN could be suitable for modeling the temporal dimension of the rainfall pattern, thus, provides better forecasting accuracy [101].

1.3.1.2. Back-Propagation Network (BPN)

Lekkas et al., 2004 have used a multilayer back propagation network and found that BPNN will not always find the correct weight and biases fort the optimum solution, whereas their results supported the hypothesis that ANNs can produce qualitative forecast. A 7 hour ahead forecast in particular proves to be of fairly high precision, especially when an error prediction technique is introduced to the ANN model [44]. In coastal areas it is an enormous prediction of tidal level, Tsong-Lin Lee, 2004, has predicted long-term tidal level using back propagation neural network, as compare to conventional harmonic method, he concluded that back-propagation neural network mode also efficiently predicts the long-term tidal levels [49]. On the basis of humidity, dew point and pressure in India, Enireddy et al., 2010, have used the back propagation neural network model for predicting the rainfall. In the training they have obtained 99.79% of accuracy and 94.28% in testing. From these results they have concluded that rainfall can predicted in future using the same method [88]. Joshi and Patel, 2011, have put in a review report on Rainfall-Runoff modeling using ANN, in the same study they have reviewed three neural network methods, Feed Forward Back Propagation (FFBP), Radial Basis Function (RBF) and Generalized Regression Neural Network (GRNN) and they have seen that GRNN flow estimation performances were close to those of the FFBP, RBF and MLR [95]. The temperature has a great effect in forecasting rainfall Amanpreet Kaur, and Harpreet Singh, 2011, have tested Artificial Neural Network in forecasting minimum temperature, they have used multi layer perceptron architecture to model the forecasting system and back propagation algorithm is used to train the network. They found that minimum temperature can be predicted with reasonable accuracy using ANN model [98]. Geetha and Selvaraj, 2011, have predicted Rainfall in Chennai using back propagation neural network model, by their research the mean monthly rainfall is predicted using ANN model. The model can perform well both in training and independent periods [102]. Predicting groundwater level is somehow a difficult task now a day because it varies place to place and round the globe, Mayilvaganan, and Naidu, 2011, have

24

attempted to forecast groundwater level of a watershed using ANN and Fuzzy Logic. A three-layer feed-forward ANN was developed using the sigmoid function and the back propagation algorithm. Now it has been observed that ANNs perform significantly better than Fuzzy Logic [105]. El-shafie et al., 2011, have tried to use neural network and regression technique for rainfall-runoff prediction finally they concluded that the results showed that the feed forward back propagation Neural Network (ANN) can describe the behaviour of rainfall-runoff relation more accurately than the classical regression model [106]. Sawaitul et al., have presented an approach for classification and prediction of future weather using back propagation algorithm, and discussed different models which were used in the past for weather forecasting, finally the study concludes that the new technology of wireless medium can be used for weather forecasting process. It also concludes that the Back Propagation Algorithm can also be applied on the weather forecasting data. Neural Networks are capable of modeling a weather forecast system [110-112].

1.4   REGION OF STUDY

Location of Chhattisgarh, India is shown in Fig. 1.1. Chhattisgarh is geographically situated in Latitude $17^0 46'$ N to $24^0 5'$N Longitude $80^0 15'$ E to $84^0 20'$ E. Ambikapur districts of Chhattisgarh, India is under consideration for the study [113] for the identification of weather prediction at district level.

1.5   DATA

1.5.1 Collection of Data

IMD data of Ambikapur station, Chhattisgarh, having maximum meteorological data (1951-2004) have been made available for the study. Data of missing years (less then 10%) have been replaced by neighboring (within two kms) station data. Even after this replacement, monthly rainfall values for few stations and for few years were still missing and those were replaced by their monthly mean values. The station averages are calculated only up to the station level. Thus, the meteorological data series so constructed are homogeneous spatially as well as temporally.

Fig. 1.2. Location of Ambikapur, Chhattisgarh, India. Geographically situated in Latitude $23^0 23'$ N to $24^0 5'$N Longitude $83^0 39'$ E elev 1958 ft.

1.5.2   Pre-Processing of Data

In this phase some pre-processing on the data has been done to choose the variables to be used, to check the relations between them and to build a regular network for the independent variables. Another important aspect according to Bryan *et al.*, [114] is to normalize the data because network-training algorithm are limited to the intervals 0 to 1 and de-normalize it after the testing phase. Finally, the cases have been randomized before splitting the data into the training, and the independent (validation) datasets.

1.5.3   Study of Data

Basu and Andhariya [115] have found that the total monsoon rainfall (TMRF) and other weather data time series behaves as chaotic series. In short, meteorological data time series (chaos) represents the following important features. For example, the weather data time series (1951-2004) of Ambikapur region in Surguja district is shown in Fig. 1.2 (a-e) representing chaotic series with the following features:

- No periodic behavior and sensitivity to initial conditions.
- Chaotic motion is difficult or impossible to forecast.
- The motion 'looks' random.
- Non-linear.

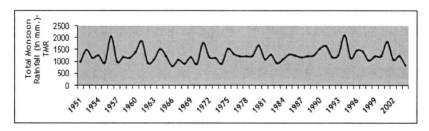

(a) TMRF data time series

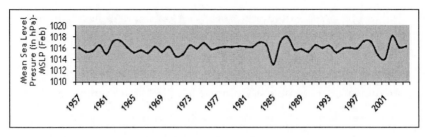

(b) Mean Sea Label Pressure (in hPa) data time series

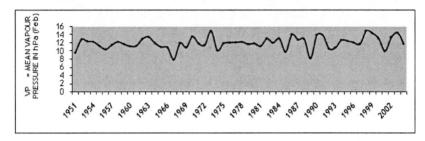

(c) Vapor Pressure (in hPa) data time series

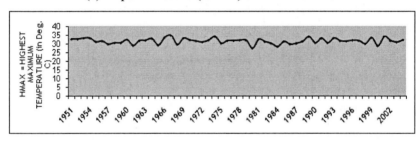

(d) Highest Maximum Temperature (in C) data time series

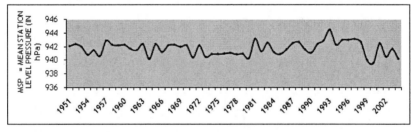

(e) Mean Station Level Pressure (in hPa) data time series

Fig. 1.2. Weather Parameter Data Time Series (1951-2004) of Ambikapur
Region Representing Chaotic Motion.

1.6   METHODOLOGY TO CHAOS PREDICTION

There are two ways by which a prediction could be made:

- Dynamical prediction – based on the physics of the controlling processes;
- Empirical/Statistical prediction – based on the statistics of the controlling processes.

The dynamical models are based on the system of nonlinear operator equations governing the atmospheric system. The physics and dynamics of the atmosphere can be better understood by none other than these set of governing equations. But in the absence of any analogue solution of this system of operator equations, numerical solutions based on approximations and assumptions are the only alternative. Furthermore, the chaotic behaviours of these nonlinear equations sensitive to initial conditions make it more difficult to solve these equations. As a result, there is limited success in forecasting the weather parameters using the numerical model. The accuracy of the models is dependent upon the initial conditions that are inherently incomplete. These systems are not able to produce satisfactory results in local and short-term cases. The performances, however, are poorer for long-range prediction of monsoon rainfall even for the larger spatial scale and particularly, for the Indian regions. As an alternative, statistical methods in which rainfall time series, treated as stochastic are widely used for long-range predication of rainfall. IMD has been using statistical models for predicting monsoon rainfall [7, 10, 115-118]. However, it is very difficult to get the same or better skill in predicting district rainfall as that of all-India monsoon rainfall using these statistical models. The neural network technique is

29

able to get rid of the above two drawbacks. Thus ANN approach as a methodology has been proposed.

1.7.   NEURAL NETWORK APPROACH TO NON-LINEAR OR CHAOS SERIES PREDICTION

Neural network technique is being applied by many scientists in the non-linear or chaos series prediction models. Neural network technique has a strong potential for pattern recognition and signal processing problems and it has the ability to predict for the future value of the time series as we discussed in literature review. This technique has successfully been applied to a variety of problems. The idea of using ANNs for forecasting is not new. The idea of using ANNs for forecasting is not new. The first application dates back to 1964. Hu, 1964 in his thesis, uses the Widrow's adaptive linear network to weather forecasting [119]. Due to the lack of a training algorithm for general multi-layer networks at the time, the research was quite limited. It is not until 1986 when the backpropagation algorithm was introduced by Rumelhart et al., 1986 that there had been much development in the use of ANNs for forecasting [120-123]. Werbos,1974, 1988 first formulates the backpropagation and finds that ANNs trained with backpropagation outperform the traditional statistical methods such as regression and Box-Jenkins approaches [124, 125]. Lapedes and Farber, 1987 conduct a simulated study and conclude that ANNs can be used for modeling and forecasting nonlinear time series [126, 127]. Elsner and Tsonis that neural network can be used successfully to predict a chaotic time series [128]. Weigend et al. ,1990, 1992; Cottrell et al.,1995 address the issue of network structure for forecasting real-world time series [129-132]. Tang et al.,1991, Sharda and Patil 1992, and Tang and Fishwick 1993, among others, report results of several forecasting comparisons between Box-Jenkins and ANN models [133,134]. Forecasting competition organized by Weigend and Gershenfeld, 1993 through the Santa Fe Institute, winners of each set of data used ANN models [135]. Research efforts on ANNs for forecasting are considerable.

1.8. CONCLUSIONS

Attempts to forecast monsoon rainfall by using statistical methods over smaller areas like a district, or monsoon periods such as a July, monsoon (June-September) remain unsuccessful as correlations fall drastically. Identification of tele-connected (i.e., predictors) data for monsoon rainfall over the district level is also a challenging task. Thus IMDs power regression model is unsuccessful especially for district or subdivision level rainfall prediction. Monsoon rainfall and other climate data time series behave as chaotic series. Chaotic motion is difficult or impossible to forecast very accurately. However, it has been found that the neural network technique has a strong potential to identify pattern as well as internal dynamics from non-linear time series data.

During an intense study of applications of various architectures of ANN, it has been found that the BPN and RBFN are the methods which have been used by most of the researchers and the result of their experiment found to be satisfactory without any scientific controversy. In general it has been observed that out of various forecasting techniques such as statistical and numerical modeling, over the meteorological data, ANN is proved to be an appropriate technique undoubtedly for

31

forecasting various weather phenomenons. It has been also proved by the contribution of Karmakar et al., 2008, 2009, that ANN is more efficient technique than others, to identify the predictors and for forecasting logn-range monsoon rainfall over the high resolution geographical region such as district or sub-division level.

This study concentrates on capabilities of ANN in prediction of several weather phenomenons such as rainfall, temperature, flood and tidal level etc. finally it has been concluded that the major architectures i.e. BPN, RBFN, MLP are sufficiently suitable to predict weather phenomenon. In the comparative study among various ANN techniques, BPN and RBFN are found as appropriate solutions for prediction of long-range weather forecasting. The study of BPN and RBFN for long range meteorological parameters pattern recognition over smaller scale geographical region shows a good performance and reasonable prediction accuracy was achieved for this model.

Chapter 2

DEVELOPMENT OF BPN MODEL USING JAVA

This chapter deals with the introduction to ANN. It mainly discusses (in detail) the development of ANN models in deterministic, n-parametric, and the training algorithm using Java.

2.1 INTRODUCTION

ANNs are predictive models closely based on the action of biological neurons. This chapter deals with the introduction of ANN (biological neural network and mathematical neural network. This chapter presented types of ANN, architecture of multilayered ANN, back-propagation learning rule, training of ANN with an example, proposed training algorithm for forecasting. It is a very challenging task to decide the architecture of ANN and its parameters. In this chapter, selection of its parameters, i.e., initial weights, learning rate (α), number of hidden layers, number of neurons in hidden layer, number of input vector in input layer, obtaining local minima and global minima have been described. It is found that, obtaining global minima during the training of ANN is a temporal timidity also has been described in this chapter. Application of ANN in deterministic forecast and parametric forecast has been described. Implementation of ANN using Java has also been described in this chapter and finally concluded the chapter.

.

2.1.1 A Brief History of ANNs

The selection of the name "Artificial Neural Network (ANN)" was one of the great pattern recognition successes of the Twentieth Century. It certainly sounds more exciting than a technical description such as "A network of weighted, additive values with nonlinear transfer functions". However, despite the name, neural networks are far from "thinking machines" or "artificial brains". A typical ANN might have hundreds of neurons. In comparison, the human nervous system is believed to have about 3×10^{10} neurons. The original "Perceptron" model was developed by Frank Rosenblatt in 1958. Rosenblatt's model consisted of three layers, (i) a "retina" that distributed inputs to the second layer, (ii) "association units" that combine the inputs with weights and trigger a threshold step function which feeds to the output layer, (iii) the output layer which combines the values. Unfortunately, the use of a step function as input to the neurons made the perceptions difficult or impossible to train. A critical analysis of perceptrons published in 1969 by Marvin Minsky and Seymore Papert pointed out a number of critical weaknesses of perceptrons, and, for a period of time, interest in perceptrons waned. Interest in ANNs was revived in 1986 when David Rumelhart, Geoffrey Hinton and Ronald Williams published "Learning Internal Representations by Error Propagation". They proposed a multi-layered ANN with nonlinear but differentiable transfer functions those avoided the pitfalls of the original perceptron's step functions. They also provided a reasonably effective training algorithm for ANNs.

2.1.2   Biological ANN

ANNs emerged after the introduction of simplified neurons by McCulloch and Pitts in 1943. These neurons were presented as models of biological neurons and as conceptual components for circuits that could perform computational tasks. The basic model of the neuron is based upon the functionality of a biological neuron. "Neurons are the basic signaling units of the nervous system" and "each neuron is a discrete cell whose several processes arise from its cell body". Fig. 2.1 (a) shows the components of neuron. Neurons and the interconnections synapses constitute the key elements for neural information processing [Fig. 2.1 (b)]. Most neurons possess tree-like structures called dendrites, which

35

receive incoming signals from other neurons across junction called synapses. Some neurons communicate with only a few nearby ones, whereas others make contact with thousands.

There are three parts in a neuron:

1. A neuron cell body.
2. Branching extensions called dendrites for receiving inputs.
3. An axon that carries the neuron's output to the dendrites of other neurons.

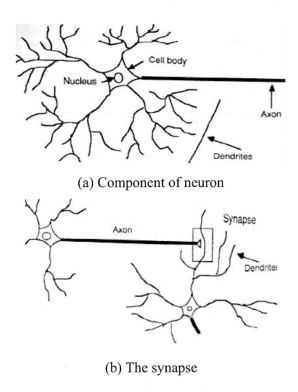

(a) Component of neuron

(b) The synapse

Fig. 2.1. Biological ANN.

2.1.3 Mathematical Model of ANN

When creating a functional model of the biological neuron, there are three basic components of importance. First, the synapses of the neuron are modeled as weights. The strength of the connection between an input and a neuron is noted by the value of the weight. Negative weight values reflect inhibitory connections, while positive values designate excitatory connections. The other two components built the model of the actual activity within the neuron cell. An adder sums up all the inputs modified by their respective weights. This activity is referred to as linear combination. Finally, an activation function controls the amplitude of the output of the neuron. An acceptable range of output is usually between 0 and 1, or -1 and 1. The mathematical ANN model is shown in Fig 2.2. The output of the neuron y_k would therefore be the outcome of some activation function on the value of v_k. From this model the interval activity of the neuron can be presented as follows:

$$v_k = \sum_{j=1}^{p} w_{kj} x_j$$

where, w_{kj} are weights and x_i are input signals/vectors.

2.1.4 Activation/Transfer Functions

Activation/Transfer functions (axons) used in the hidden layer (s), which are shown in the following Table 2.1 and Fig. 2.3. As mentioned above, the activation function acts as a function, such that the

37

output of a neuron in a ANN is between certain values (usually 0 and 1, or -1 and 1). Activation functions are denoted by Φ (.) or f (.). The first five functions (Table 2.1) are non linear. If a non-linear function is used one has to normalize the data, before processing it, in the neural system itself, in order to fit within the output range. In the output layer the transfer function, dealing with regression, the desired response is a continuous function of the input, one will use an axon with an infinite output range (bias axon or linear axon).

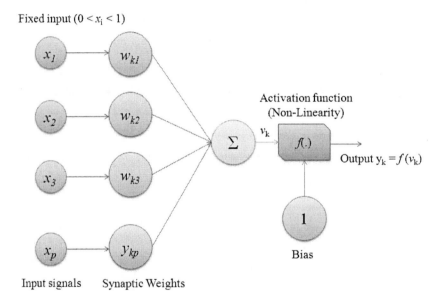

Fig. 2.2. Mathematical ANN Model.

Table 2.1. Activation/Transfer Functions

Axon	Output Range	Features
Tanh Axon	-1 to 1	Nonlinear axon of choice.
Sigmoid Axon	0 to 1	Same general shape as TanhAxon.
Linear Tanh Axon	-1 to 1	Piecewise linear approximation to TanhAxon.
Linear Sigmoid Axon	0 to 1	Piecewise linear approximation to SigmoidAxon.
SoftMax Axon	-0 to 1	Outputs sum to 1. Used for classification.
Bias Axon	Infinite	Linear axon with adjustable slope and adaptable bias.
Linear Axon	Infinite	Linear axon with adaptable bias.

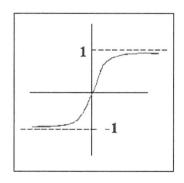

(a) Tanh Axon.

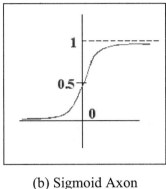

(b) Sigmoid Axon

Fig. 2.3. Activation Functions.

2.1.5   Types of ANNs

When used without qualification, the terms Neural Network (NN) and ANN usually refer to a multi-layer perceptron feed forward back-propagation network. However, there are many other types of ANN such as Counter Propagation Neural Networks (CPNN), Radial Basis Function Networks (RBFN), Cascade Correlation Neural Network (CCNN), Probabilistic Neural Network (PNN) and General Regression Neural Networks (GPRN), etc., descried in Table 2.2. As per the objectives of the work any of them can be used. However in this study, Multi-Layered Back Propagation (BPN) has been proposed because for three reasons. First, a lot of studies have been made (Section 1.3) on BPN and their abilities, especially in the case of chaos prediction. Second, the availability of meteorological data corresponds to the needs of BPN with input and output data. And third, its use and construction are easy. In this study, BPN is implemented by using Java technology for deterministic, n-parameter forecast and spatial interpolation of climate variable based on geo-coordinate.

Table 2.2. Types of Neural Networks

Type	Description	Primary Use or Advantage
CPNN	Developed by Robert Hecht Nielson is beyond the representation limits of single layer networks. This is multi-layer perceptron network (MLP) based on various combining structure of input, clustering and output layers. Compared to the BPN it reduces the time by one hundred times. CPNN different from BPN, in the sense that it provides solution for those applications, which cannot have larger iterations (epochs).	CPNN can be used for data compression, approximation of functions, pattern association, pattern completion and signal enhancement applications.
RBFN	This is multi-layer perceptron network (MLP) based on Gaussian Potential functions. There exist 'n' number of input neurons and 'm' number of output neurons with the hidden layer existing between input and output layer. The interconnection between input and output layer forms hypothetical connection and between the hidden	RBFN can be used for data compression, approximation of functions, and recognizing pattern.

	and output layer forms weighted connections.	
CCNN	Cascade correlation neural networks (Fahlman and Libiere, 1990) are self organizing networks. The network begins with only input and output neurons. During the training process, neurons are selected from a pool of candidates and added to the hidden layer	CCNN can be used recognizing pattern.
PNN	PNN is based on the theory of Bayesian classification and estimation of probability density functions. It is necessary to classify the input vectors into one of the two classes in a Bayesian optimal manner.	PPN can be used recognizing pattern
GRNN	Probabilistic and General Regression Neural Networks have similar architectures, but there is a fundamental difference: Probabilistic networks perform classification where the target variable is categorical, whereas general regression neural networks perform regression where the target variable is continuous	GRNN can be used recognizing pattern.

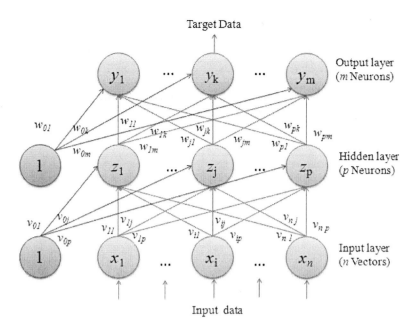

Fig.2.4. Architecture of BPN Model.

2.2 ARCHITECTURE OF BPN MODEL

A BPN with one layer of p-hidden units is shown in the Fig. 2.4. The y output unit has W_{0k} bias and p-hidden unit has V_{0k} as bias. It is found that both the output units and hidden units have bias. The bias acts like weights on connections from units whose output is always 1. From Fig. 2.4, it is clear that the network has one input layer, one hidden layer and one output layer. There can be any number of hidden layers. The input layer is connected to the hidden layer and the hidden layer is connected to the output layer by means of interconnected weights. In Fig. 2.4, only the feed forward phase of operation is shown. But during the back-propagation phase of learning, the signals are sent in the reverse direction. The increase in number of hidden layers results in

43

computational complexity of the network. As a result, the time taken for convergence and to minimize the error may be very high.

2.3   BACK PROPAGATION NETWORK (BPN)

Back-propagation introduced by Rumelhart *et al.*,1986 is a systematic method for training multi-layer ANN [120-123]. It has a mathematical foundation. It is a multi-layer forward network using extend gradient-descent based delta-learning rule commonly known as back propagation (of errors) rule. Back-propagation provides a computationally efficient method for changing the weights in a feed forward network, with differentiable activation function units, to learn a training set of input-output example. Being a gradient descent method, it minimizes the MSE of the output computed by the network. The network is trained by supervised learning method. Supervised training is the process of providing the network with a series of sample inputs and comparing the output with the expected response. The aim of this network is to train the network in order to achieve a balance between the ability to respond correctly to the input patterns that are used for training and to provide good responses to the inputs that are similar [137].

2.4   GENERALIZED DELTA LEARNING (GDL) RULE OR BACK PROPAGATION (BP) RULE

The total square error of the output computed by network is minimized by a gradient descent method known as back propagation or generalized delta learning rule [138]. Consider an arbitrary activation function $f(x)$. The derivation of the activation function is denoted by $f'(x)$.

Parameters

The various parameters used in the back propagation rule and the training algorithm are as follows:

Input training vector x	=	$(x_1...x_i...x_n)$
Output target vector t	=	$(t_1...t_k...t_m)$
δ_k	=	Error at output unit y_k.
α	=	Learning rate.
V_{oj}	=	Bias on hidden unit j.
Z_j	=	Hidden unit j.
w_{ok}	=	Bias on output unit k.
y_k	=	Output unit k.

Let

$$y_{-ink} = \sum_i z_i w_{jk}$$

$$z_{-inJ} = \sum_i v_{ij} x_i$$

$$y_k = f(y_{-ink})$$

The error to be minimized is

$$E = 0.5 \sum_k [t_k - y_k]^2$$

By use of chain rule

$$\frac{\partial E}{\partial w_{jk}} = \frac{\partial}{\partial w_{jk}} \left(0.5 \sum_k [t_k - y_k]^2\right)$$

$$= \frac{\partial}{\partial w_{jk}} \left(0.5 \sum_k [t_k - f(y_{-ink})]^2 \right)$$

$$= -[t_k - y_k] \frac{\partial}{\partial w_{jk}} \left(f(y_{-ink}) \right)$$

$$= -[t_k - y_k] \left(f(y_{-ink}) \right) \frac{\partial}{\partial w_{jk}} \left(y_{-ink} \right)$$

$$= -[t_k - y_k] \left(f'(y_{-ink}) \right) \frac{\partial}{\partial w_{jk}} \left(y_{-ink} \right)$$

$$= -[t_k - y_k] \left(f'(y_{-ink}) \right) z_j$$

Let us define

$$\delta_k = [t_k - y_k] f'(y_{-ink})$$

Weights are connections to the hidden unit z_j

$$\frac{\partial E}{\partial v_{jk}} = -\sum_k [t_k - y_k] \frac{\partial}{\partial v_{ij}} y_k$$

$$\frac{\partial E}{\partial v_{jk}} = -\sum_k [t_k - y_k] f(y_{ink}) \frac{\partial}{\partial v_{ij}} y_{-ink}$$

$$\frac{\partial E}{\partial v_{jk}} = -\sum_k \delta_k \frac{\partial}{\partial v_{ij}} y_{-ink}$$

Rewriting the equation and substituting the value of y_{-ink}

$$\frac{\partial E}{\partial v_{jk}} = -\sum_k \delta_k \frac{\partial}{\partial v_{ij}} \left(\sum z_j - w_{jk} \right)$$

$$\frac{\partial E}{\partial v_{jk}} = -\sum_k \delta_k w_{jk} \frac{\partial}{\partial v_{ij}} z_j$$

$$\frac{\partial E}{\partial v_{jk}} = -\sum_k \delta_k w_{jk} \frac{\partial}{\partial v_{ij}} f(z_{inj})$$

$$\frac{\partial E}{\partial v_{jk}} = -\sum_k \delta_k w_{jk} \frac{\partial}{\partial v_{ij}} \ f'(z_{inj})(x_i)$$

$$\delta_j = -\sum_k \delta_k w_{jk} \frac{\partial}{\partial v_{ij}} \ f'(z_{inj})$$

The weight updation for output unit is given by

$$\Delta w_{jk} = -\alpha \frac{\partial E}{\partial w_{jk}}$$

$$\Delta w_{jk} = [t_k - y_k] f'(y_{-ink}) z_j$$

$$\Delta w_{jk} = \alpha \delta_k z_j$$

The weight updating for the hidden unit is given by

$$\Delta v_{jk} = -\alpha \frac{\partial E}{\partial v_{ij}}$$

$$\Delta v_{jk} = \alpha f'(z_{-inj}) x_i \sum_k \delta_k w_{jk}$$

$$\Delta v_{jk} = \alpha \delta_j x_i$$

This is the generalized delta rule used in the back propagation network during the training.

2.5   TRAINING THE NEURAL NETWORK

The training algorithm of back propagation involves four stages [137,138], viz.

1. Initialization of the weights
2. Feed forward
3. Back-propagation error
4. Updation of the weights and biases.

During the first stage, which is the initialization of weights, some small random values are assigned. During feed forward stage each input unit (x_i) receives an input signal and transmits this signal to each of hidden units $z_i...z_p$. Each hidden unit then calculates the activation function and sends its z_j to output unit. The output unit calculates the activation function to form the response of the network for the given input pattern.

During the back-propagation of errors, each output unit compares its computed activation y_k with its target t_k to determine the associated error for that pattern with the unit. Based on the error, the factor $\delta_k (k = 1,..., m)$ is compared and is used to distribute the error at output unit y_k based to all units in the previous layer. Similarly the factor $\delta_j (j = 1,..., p)$ is computed for each hidden unit z_j.

During the final stage, the weight and biases are updated using the δ factor and the activation. The training algorithm used in the back propagation network and its various steps are as follows:

Initialization of Weights

Step 1: Initialize weights to small random values.

Step 2: While stopping condition is false, perform Steps 3 to10, otherwise stop.

Step 3: For each training pair, execute Steps 4-9.

Feed Forward

Step 4: Each input unit receives the input signal x_i and transmits this signals to all units

48

in the layer above i.e., hidden units.

Step 5: Each hidden unit $(z_j, j = 1...p)$ sums its weighted input signals

$$z_{-inj} = v_{oj} + \sum_{i=1}^{n} x_i v_{ij}$$

applying activation function

$$Z_j = f(z_{inj})$$

and sends this signal to all units in the layer above i.e., output units.

Step 6: Each output unit $(y_k, k = 1...m)$ sum its weighted input signals.

$$y_{-ink} = w_{ok} + \sum_{j=1}^{p} z_j w_{jk}$$

applying activation function

$$y_k = f(y_{-ink})$$

Back Propagation Errors

Step 7: Each output unit $(y_k, k = 1...m)$ receives a target pattern corresponding to an input pattern error information term is calculated as

$$\delta_k = (t_k - y_k) f(y_{-ink})$$

Step 8: Each hidden unit $(z_j, j = 1...n)$ sums its delta inputs from units in the layer above

$$\delta_{-inj} = \sum_{k=1}^{m} \delta_j w_{jk}$$

The error information term is calculated as

$$\delta_j = \delta_{-inj} f(z_{-inj})$$

Updation of Weights and Biases

Step 9: Each output unit $(y_k, k = 1...m)$ updates its bias and weights $(j = 0...p)$

The weights correction term is given by

$$\Delta W_{jk} = \alpha \delta_k z_j$$

And the bias correction term is given by

$$\Delta W_{ok} = \alpha \delta_k$$

Therefore,

$$W_{jk}(new) = W_{jk}(old) + \Delta W_{jk}, W_{ok}(new) = W_{ok}(old) + \Delta W_{ok}$$

Each hidden unit $(z_j, j = 1...p)$ updates its bias and weights $(i = 0...n)$ the weights correction term

$$\Delta V_{ij} = \alpha \delta_j x_i$$

The bias correction term

$$\Delta V_{oj} = \alpha \delta_j$$

therefore,

$$V_{ij} = V_{ij}(old) + \Delta V_{ij}, V_{oj}(new) = V_{oj}(old) + \Delta V_{oj}$$

Step 10: Test the stopping condition. The stopping condition may be minimizing of errors, number of epochs etc.

Example: Finding the new weights when the network is presented the input pattern [0.6   0.8 0] and target output is 0.9. Using learning rate $\alpha = 0.3$ and binary sigmoid activation function.

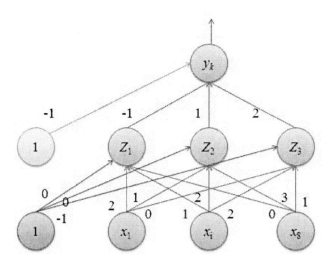

Fig. A back propagation neural network.

Solution: The solution to the problem is as follows:

Step 1: Initialize the weight and bias

$$W = [-1 \quad 1 \quad 2], \qquad W_0 = [-1]$$

Step 2:

$$V = \begin{matrix} 2 & 1 & 0 \\ 1 & 2 & 2 \\ 0 & 3 & 1 \end{matrix} \qquad V_0 = [0 \quad 0 \quad -1]$$

Step 3: For each training pair $\quad X = [0.6 \quad 0.8 \quad 0] \qquad t = [0.9]$

Feed Forward Stage

Step 4: $$Z_{-inj} = V_{0j} + \sum_{i=1}^{n} (x_i V_{ij})$$

$$Z_{-in1} = V_{01} + \sum_{i=1}^{3} (x_i V_{i1})$$

$\Rightarrow V_{01} + x_1 V_{11} + x_2 V_{21} + x_3 V_{31}$

$\Rightarrow 0 + 0.6 \times 2 \quad + 0.8 \times 1 + 0 \times 0$

$\Rightarrow 1.2 + 0.8 = 2$

51

$$Z_{-in2} = V_{02} + \sum_{i=1}^{3} (x_i \, V_{i2})$$

$\Rightarrow V_{02} + x_1 \, V_{12} + x_2 \, V_{22} + x_3 \, V_{32}$

$\Rightarrow 0 + 0.6 \times 1 \quad + 0.8 \times 2 + 0 \times 3$

$\Rightarrow 0.6 + 1.6 \quad = \quad 2.2$

$$Z_{-in3} = V_{03} + \sum_{i=1}^{3} (x_i \, V_{i3})$$

$\Rightarrow V_{03} + x_1 \, V_{13} + x_2 \, V_{23} + x_3 \, V_{33}$

$\Rightarrow -1 + 0.6 \times 0 \quad + 0.8 \times 2 + 0 \times 1$

$\Rightarrow -1 + 1.6 \quad = \quad 0.6$

$Z_1 = f(Z_{-in1}) = \frac{1}{1+e^{-2}} \quad = \quad 0.8808$

$Z_1 = f(Z_{-in2}) = \frac{1}{1+e^{-2.2}} \quad = \quad 0.9002$

$Z_1 = f(Z_{-in3}) = \frac{1}{1+e^{-0.6}} \quad = \quad 0.646$

Step 5: Calculate Y_{-ink}

$$Y_{-ink} = W_{0k} + \sum_{j=1}^{p} (Zj \, W_{jk})$$

$$Y_{-in1} = W_{01} + \sum_{j=1}^{3} (Zj \, W_{j1})$$

$\Rightarrow W_{01} + Z_1 \, W_{11} + Z_2 \, W_{21} + Z_3 \, W_{31}$

$\Rightarrow -1 + 0.8808 \times -1 + 0.9002 \times 1 + 0.646 \times 2$

$\Rightarrow -1 - 0.8808 + 0.9002 + 1.292 \quad\quad = \quad 0.3114$

Calculate the output signal

$Y_1 = f(Y_{-in1}) = \frac{1}{1+e^{-0.3114}} \quad = \quad 0.5772$

Back propagation of error

Step 6: Calculating error information term δ_k

$\delta_k = (t_k - Y_k) f^1 (Y_{-ink})$

$\delta_1 = (t_1 - Y_1) f^1 (Y_{-in1})$

We know that for binary sigmoid function

$f^1(x) = f(x) (1-f(x))$

$f^1 (Y_{-in1}) = f(Y_{-in1}) (1 - f(Y_{-in1}))$

$= 0.5772 (1 - 0.5772)$

$= 0.2440$

$\delta_1 = (t_1 - Y_1) f^1 (Y_{-in1})$

$= (0.9 - 0.5772)(0.2440)$

$= 0.0788$

Step 7: Back Propagation to be first hidden layer (I = 1, 2, 3) we have to calculate δ_{-in1}

$$\delta_{-inj} = \sum_{k=1}^{m} \left(\delta_k W_{jk} \right)$$

$$\delta_{-inj} = \sum_{k=1} \left(\delta_k W_{jk} \right)$$

$\delta_{-in1} = \delta_1 W_{11} = 0.0788 \times -1 = -0.0788$

$\delta_{-in2} = \delta_1 W_{21} = 0.0788 \times 1 = 0.0788$

$\delta_{-in3} = \delta_1 W_{31} = 0.0788 \times 2 = 0.01576$

To calculate error term in hidden layer

$\Delta = \delta_{-inj} f^1 (Z_{-inj})$

$f^1 (Z_{-in1}) = f(Z_{-in1}) (1 - f(Z_{-in1}))$

$= 0.8808 (1 - 0.8808) = 0.1049$

53

$$\Delta_1 = \delta_{-in1}\, f(Z_{-in1})(1 - f(Z_{-in1}))$$
$$= (-0.0788)(0.1049) \qquad = -0.0083$$

$$\Delta_2 = \delta_{-in2}\, f^1\,(Z_{-in2})$$
$$f^1\,(Z_{-in2}) = f(Z_{-in2})\,(\,1 - f(Z_{-in2}))$$
$$= (0.9002)(1-0.9002) \qquad = 0.09$$

$$\Delta_2 = 0.0788 \times 0.09 = 0.0071$$

$$\Delta_3 = \delta_{-in3}\, f^1\,(Z_{-in3})$$
$$f^1\,(Z_{-in3}) = f(Z_{-in3})\,(\,1 - f(Z_{-in3}))$$
$$= (0.646)(1 - 0.646) \qquad = 0.2286$$

$$\Delta_3 = 0.1576 \times 0.2286 = 0.0361$$

Step 8: Weight updation

$$\Delta V_{ij} = \alpha\, \Delta_j\, x_i$$
$$x = [\,0.6 \quad 0.8 \quad 0\,]$$
$$\Delta = [\,-0.0083 \quad 0.0071 \quad 0.0361\,]$$
$$\alpha = 0.3$$

$$\Delta V_{11} = \alpha\, \Delta_1\, x_1 \;\; = 0.3 \times -0.0083 \times 0.6 \quad = -0.0015$$
$$\Delta V_{12} = \alpha\, \Delta_2\, x_1 \;\; = 0.3 \times 0.0071 \times 0.6 \quad = 0.0013$$
$$\Delta V_{21} = \alpha\, \Delta_1\, x_2 \;\; = 0.3 \times -0.0083 \times 0.8 \quad = -0.002$$
$$\Delta V_{22} = \alpha\, \Delta_2\, x_2 \;\; = 0.3 \times 0.0071 \times 0.8 \quad = 0.0017$$
$$\Delta V_{13} = \alpha\, \Delta_3\, x_1 \;\; = 0.3 \times 0.0361 \times 0.6 \quad = 0.0065$$
$$\Delta V_{23} = \alpha\, \Delta_3\, x_2 \;\; = 0.0087$$
$$\Delta V_{31} = \Delta V_{32} = \Delta V_{33} = 0$$
$$\Delta V_{01} = \alpha\, \Delta_1 \qquad ; \qquad \Delta V_{02} = \alpha\, \Delta_2 \qquad ; \qquad \Delta V_{03} = \alpha\, \Delta_3$$
$$\Delta V_{01} = 0.3 \times -0.0083 \qquad = \qquad -0.0025$$
$$\Delta V_{02} = 0.3 \times 0.0071 \qquad = \qquad 0.0021$$
$$\Delta V_{03} = 0.3 \times 0.0361 \qquad = \qquad 0.0108$$

$V_{new} = V_{old} + \Delta V_1$

$V_{11(new)} = V_{11(old)} + \Delta V_{11} = 2 - 0.0015 = \qquad 1.9985$

$V_{12(new)} = V_{12(old)} + \Delta V_{12} = 1 + 0.0013 = \qquad 1.0013$

$V_{13(new)} = V_{13(old)} + \Delta V_{13} = 0 + 0.065 = \qquad 0.065$

$V_{21(new)} = V_{21(old)} + \Delta V_{21} = 1 - 0.002 = \qquad 0.998$

$V_{22(new)} = V_{22(old)} + \Delta V_{22} = 2 + 0.0017 = \qquad 2.0017$

$V_{23(new)} = V_{23(old)} + \Delta V_{23} = 2 + 0.0087 = \qquad 2.0087$

$V_{31(new)} = V_{31(old)} + \Delta V_{31} = 0 + 0 = \qquad 0$

$V_{32(new)} = V_{32(old)} + \Delta V_{32} = 3 + 0 = \qquad 3$

$V_{33(new)} = V_{33(old)} + \Delta V_{33} = 1 + 0 = \qquad 1$

So
$$V = \begin{matrix} 1.9985 & 1.0013 & 0.085 \\ 0.998 & 2.0017 & 2.0087 \\ 0 & 3 & 1 \end{matrix}$$

$\Delta W_{0k} = \alpha\, \delta_k\, Z_j$

$\Delta W_{11} = \alpha\, \delta_1\, Z_1 = \qquad 0.3 \times 0.0788 \times 0.8808 = \qquad 0.0208$

$\Delta W_{12} = \alpha\, \delta_1\, Z_2 = \qquad 0.3 \times 0.0788 \times 0.9002 = \qquad 0.0212$

$\Delta W_{13} = \alpha\, \delta_1\, Z_3 = \qquad 0.3 \times 0.0788 \times 0.646 = \qquad 0.0153$

$\Delta W_{11(new)} = W_{11(old)} + \Delta W_{11} = \qquad -1 + 0.0208 = \qquad -0.9792$

$\Delta W_{12(new)} = W_{12(old)} + \Delta W_{12} = \qquad 1 + 0.0212 = \qquad 1.0212$

$\Delta W_{13(new)} = W_{13(old)} + \Delta W_{13} = \qquad 2 + 0.0153 = \qquad 2.0153$

$\Delta W_0 = \alpha\, \delta_1 = \qquad 0.3 \times 0.0788 = \qquad 0.02364$

So $W = [0.9792 \quad 1.0212 \quad 2.0153]$

Thus the weights are calculated. The process can be continued up to any specified stopping condition. We can understand the Back-propagation algorithm through further subsequent straightforward approach. For illustration, think about following neural network as depicted in Figure

2.5 wherein two input vectors x_1 and x_2, two neurons in hidden layer with one neuron in output layer is used.

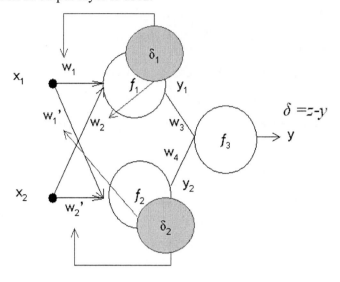

Fig. 2.5. Two Input Vectors and Two Hidden Neurons ANN.

Feed forward process:

Calculate $e_1 = x_1 w_1 + x_2 w_2$ and $e_2 = x_1 w_1' + x_2 w_2'$

Calculate $y_1 = f_1(e_1)$ and $y_2 = f_2(e_2)$

Calculate $y = f_3(e_1 w_3 + e_2 w_4)$

Calculate $\delta = z - y$.

Back propagation to optimize weight:

Calculate $\delta_1 = w_3 \delta$ and $\delta_2 = w_4 \delta$

then set new weights as follows

$$New(w_1) = w_1 + \eta \delta_1 \frac{\partial}{\partial e} f_1(e_1) x_1$$

$$New\left(w_1'\right) = w_1' + \eta\delta_1 \frac{\partial}{\partial e} f_1(e_1)x_2$$

$$New\left(w_2\right) = w_2 + \eta\delta_2 \frac{\partial}{\partial e} f_2(e_2)x_1$$

$$New\left(w_2'\right) = w_2' + \eta\delta_2 \frac{\partial}{\partial e} f_1(e_2)x_2$$

Parameter η affects network training speed. There are few techniques to select this parameter. The first method is to start training process with large value of the parameter between close interval [0 1]. While weights coefficients are being established the parameter is being decreased gradually. The second, more complicated method, starts training with small parameter between close interval [0 1]. During the training process the parameter is being increased when the training advances and then decreased in the final stage.

2.6   PROPOSED BPN MODEL

The following diagram (Fig. 2.6) illustrates a Three Layered BPN. This network has input layer (at the bottom), one hidden layer (at the middle) and output layer (on the top). The model as shown in Fig. 2.6, has n input vectors at the input layer $(x_1...x_i...x_n)$. p neurons in hidden layer $(z_1...z_j...z_p)$ and one neuron (y_k) in output unit to observe desired meteorological data as target variable. Bias on hidden unit j i.e., V_{oj} and bias on output unit k i.e., w_{ok} with $n \times p + p$ trainable weights are used in the network. Output target value for the network is actual data to be predicted.

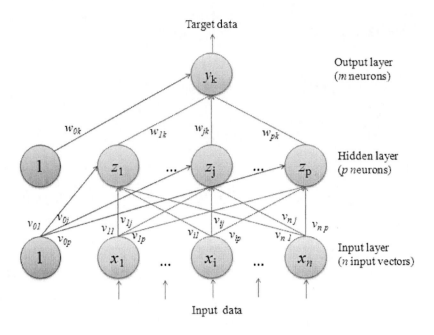

Fig. 2.6. Architecture of BPN Model.

The neurons output can be obtained as $f(x_j)$ where, f is a transfer function (axon), typically the sigmoid (logistic or tangent hyperbolic) function. Sigmoid function $f(x) = \dfrac{1}{1 + e^{-\delta x + \eta}}$, where δ determines the slope and η is the threshold. In the proposed model $\delta = 1$, $\eta = 0$ are considered such that $\forall \pm n \in I^+$, the output of the neuron will be in close interval [0, 1] as shown in Fig. 2.7. The model performance is determined by comparison between MAD (% of mean) and SD (% of mean) as shown in Equation 2.2 and 2.3 respectively. Where x_i are random variable with mean μ, and p_i s is predicted value.

$$MAD = \left| \frac{1}{n} \sum_{i=1}^{n} (x_i - p_i) \right|$$

58

$$SD = \sqrt{\frac{1}{n} \sum_{i=1}^{n} (x_i - \mu)^2}$$

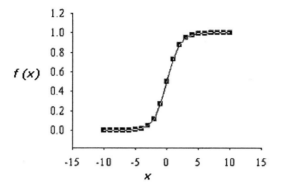

Fig. 2.7. Output of Sigmoid Axon.

2.7   PROPOSED TRAINING ALGORITHM

The proposed algorithm to train the deterministic ANN model as well as n-parameter ANN model is descried as below.

Algorithm 2.1. Training Algorithm of Deterministic ANN Model.

Set initial weights w_k such that $(0 \prec w_k \prec 1)$.

Consider training period: m;

Set number of input vector ρ such that $\rho \prec m$.

do{

$$for(i = 1; i \leq (m - \rho); i + +)$$

{

$$for(j = 1; j \leq= (\rho + j); j + +)$$

{

        input vector X_j ;

}

Feed Forward the Network and calculate target/output variable y; by using Rumelhart *et al.*, 1986 BPN algorithm ;

Calculate error factor $\delta = actual(z)$ - target (y);

Initialize predicted value $P_{\rho+i}$ = target (y);

Back propagate and calculate new weights w_k by using BPN algorithm;

}

Calculate MSE between series X_i and P_i ;

$$MSE = \left| \frac{1}{m} \left(\sum_{i=1}^{m} X_i - P_i \right) \right|$$

set $i = 1; j = 1$;

} $while\big(MSE \prec 0.000001\big);$

Algorithm 2.2. Training Algorithm of N-Parameter ANN Model

Set initial weights w_k such that $\big(0 \prec w_k \prec 1\big).$

Consider training period :m;

do{

       $for\big(i = 1; i \leq m; i + +\big)$

       {

              $for\big(j = 1; j \mathrel{<}= m; j + +\big)$

{

    //Read N-Parameter

$$X1_j; X2_j...Xn_j;$$

}

Feed Forward the Network and calculate target/output variable y; by using BPN algorithm;

Calculate error factor $\delta = actual(z)$ - target (y);

Initialize predicted value P_i = target (y);

Back propagate and calculate new weights w_k by using Rumelhart *et al.*, 1986 BPN algorithm;

}

Calculate MSE between series X_i and P_i;

$$MSE = \left| \frac{1}{m} \left(\sum_{i=1}^{m} X_i - P_i \right) \right|$$

set $i = 1; j = 1$;

} $while(MSE \prec 0.000001)$;

2.8   SELECTION OF PARAMETERS

For the efficient operation of BPN, it is necessary for the appropriate selection of the parameters used for network. The selections of parameters are discussed in detail in this section.

2.8.1.   Initial Weights

It will influence whether the network reaches a global (or only a local) minima of the error and if so how rapidly it converges. If initial weight is too large the initial input signals to each hidden or output unit will fall in the saturation region where the derivative of sigmoid has very small value (f (network) = 0). If initial weights are too small, the network output to a hidden or output unit will approach zero, which then causes extremely slow learning. To get the best result the initial weights (and biases) are set to random number between 0 and 1, -0.5 and 0.5 or between –1 and 1. The initial weights (bias) can be done randomly and there is also a specific approach. The faster learning of a BPN can be obtained by using Nguyen-Widrow (NW) initialization. This method is designed to improve the learning ability of the hidden units.

$$\beta = 0.7(p)^{1/n}$$

where, n = number of input units, p = number of hidden units, β = scale factor.

2.8.2 Learning Rate (α)

A high learning rate leads to rapid learning but the weights may oscillate, while a lower learning rate leads to slower learning. Methods suggested for adopting learning rate are as follows:

1. Start with high learning rate between 0 and 1 and steadily decrease it. Changes in the weight vector must be small in order to reduce oscillation or any divergence.
2. A simple suggestion is to increase the learning rate in order to improve performance and to decrease the learning rate in order to worsen the performance.

3. Another method is to double the learning rate until the error value worsens.

In this study, $\alpha = 0.3$ is selected for nearly all of the cases.

2.8.3 Number of Hidden Layers

For nearly all problems, one hidden layer is sufficient. Using two hidden layers rarely improves the model, and it may introduce a greater risk of converging to a local minima. There is no theoretical reason for using more than two hidden layers [139].

2.8.4 Number of Neurons in Hidden Layers (*p*)

The model has been examined with districts and subdivisions TMRF time series [136-139]. It has been found that, if number of neurons (*p*) increases in the hidden layer, the MAD (% of mean) between actual and predicted value increases. In other words, the relation between hidden neurons (*p*) and MAD (% of mean) is shown as –

$$MAD \propto p$$
$$MAD = cp$$

where, c is error constant. The relationship between number of neurons (*p*) in hidden layer and MAD (% of mean) is shown in Fig. 2.8 (a) and Table 2.3. Relationship between error constant (c) and number of neurons (*p*) in hidden layer is shown in (b) and Table 2.3. Thus *p* = 2 or 3 (depending on internal dynamics of time series) is sufficient for practically all cases.

2.8.5 Number of Input Vectors in Input Layer (n)

For all cases, the relation between number of input (n) and MAD (% of mean) has been shown in Fig. 2.9 and in Table 2.4. It is observed that MAD (% of mean) is inversely proportional to n [136-139]. Thus n = 11 or more (depending on internal dynamics of time series) input vectors have been selected for nearly all cases.

$$MAD \propto \frac{1}{n}$$

2.9   LOCAL MINIMA AND GLOBAL MINIMA

Consider minimizing process (epoch) of MSE during training period of BPN model in deterministic forecast for long range TMRF data time series of Ambikapur region as shown in Fig. 2.10. The training started with initial set of weights between 0 and 1 at point 'P' using algorithm 2.1 (Chapter 2). Minimum MSE = 2.95E-04 is encountered at point M_L after 540000 epochs. M_L is called local minima. After 7500000 epochs, the MSE becomes 2.16E-04 marked by the point M_G (Fig. 2.10). This point is called global minima. After this point MSE is exhibiting an increasing trend and is considered as over training of the network. Identification of these two points and training process is an effort. No authors have found in the literature till date, who evidently mentioned these facts in his contributions. Obtaining global minima where we may

64

stop training process really-really is a temporal timidity, and some time it may be limitations of ANN model. Some authors already declared that the training of the network is a very difficult task may be not trained at all. In this study network has been trained successfully. Thus, this experimental result is extremely useful for the readers.

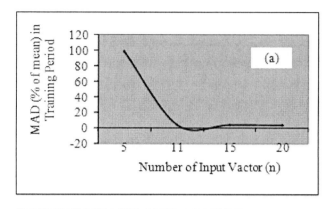

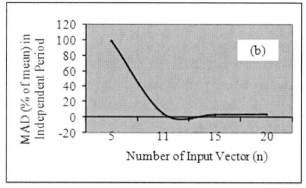

Fig. 2.9. Relation between n and MAD (% of mean) (a) In independent. period (b) In training period

65

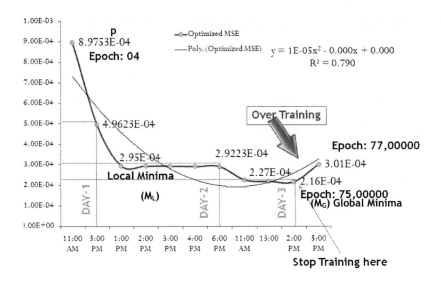

Fig. 2.10. Minimizing Error: Local Minima and Global Minima (MSE).

2.10 ARCHITECTURE OF BPN IN DETERMINISTIC FORECAST

In deterministic forecast, model uses data of the past years to forecast future. Immediate past n years of data are used as input to predict the $(n+1)^{th}$ year meteorological data. The architecture of model is shown in following Fig. 2.11. Where, n vectors in input layer to input past n years data, $n \times p + p + p + 1$ trainable weights including biases, p Neurons in hidden layer, and one neuron in output layer used to observe $(n+1)^{th}$ year data. The hidden neurons output is obtained as sigmoid function $f(x) = \dfrac{1}{1+e^{-\delta x+\eta}}$. The proposed simulator is shown in the Fig. 2.14.

66

2.11   ARCHITECTURE OF BPN IN N-PARAMETER FORECAST

In parametric forecast, parameters or predictors (i.e., meteorological data which are physically connected with targeted meteorological data) data time series of each year are used as input in the model to predict targeted meteorological data. The architecture of model is shown in following Fig. 2.12. Where, n vectors in input layer to input n parameters, $n \times p + p + p + 1$ trainable weights including biases, p neurons in hidden layer, and one neuron in output layer used to observe targeted data. The neurons output is obtained as sigmoid function $f(x) = \dfrac{1}{1 + e^{-\delta x + \eta}}$. The proposed simulator is shown in the Fig. 2.14.

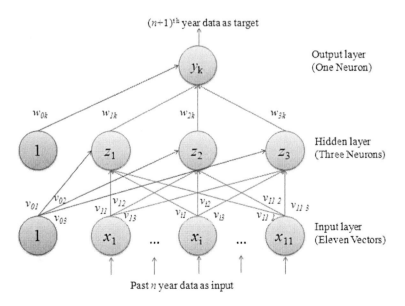

Fig. 2.11. Architecture of BPN Model in Deterministic Forecast.

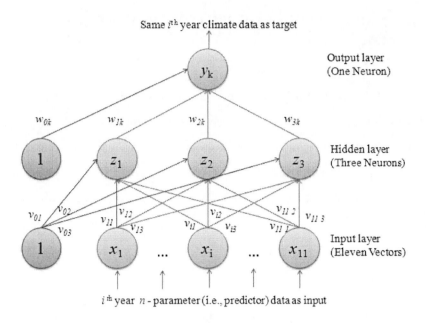

Fig. 2.12. Architecture of *n*-parameter BPN Model.

2.13   DEVELOPMENT OF BPN MODEL USING JAVA

To develop BPN in deterministic, n-parametric, forecast over the districts as well as subdivisions level and spatial interpolation of mean climate variable of Chhattisgarh, the proposed algorithm described in 2.1 and 2.2 have been implemented by using Java technology. Evaluated code in terms of Java methods are not given in the book. We will release code in the next edition. The method's prototype and their description are described in the following Table 2.5. Interfaces of developed BPN in forecast and interpolation are shown in Fig. 2.14 (a-h) and Fig. 2.15 (a-f) respectively.

Table 2.5. Designed Method's and Their Description

No.	Method	Description
1.	public void getInitWeights()	Used to initialize random weights and used to call method processNetwork().
2.	public void getInputIndData()	Used to input vector in independent period from database.
3.	public void processNetwork()	Used to calculate MSE. Used to call traniedNetwork() under the condition MSE<0.00001.
4.	public void traniedNetwork()	Used to train the network. Initially it will call method inputData() to input time series then after it will call getTargetData() method to obtain output of the network and finally will call updateWeights() to update weights by Rumelhart *et al.*, 1986back-propagation technique [136].
5.	public void inputData()	Used to input time series from database in training period
6.	public void getTargetData()	Used to obtain target data i.e., output of the network.
7.	public void updateWeights()	Used to update network by Rumelhart *et al.*, back-propagation technique [136]
8.	public void getValidationOfThe Model()	Used to identify the performance of the network.

69

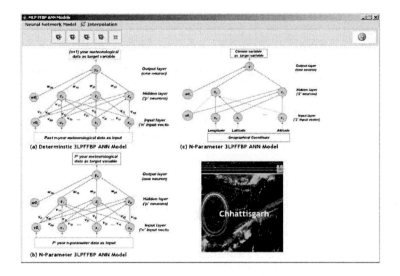

(a) BPN Model in deterministic and parametric forecast

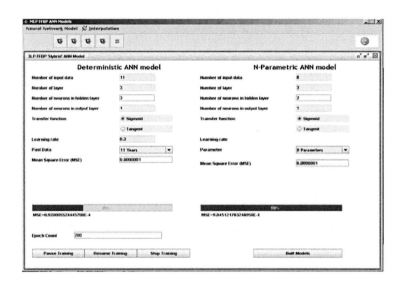

(b) Parameter selection frame of BPN in deterministic and parametric forecast

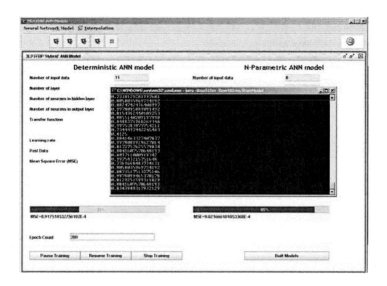

(c) Internal Parallel Processing Window During Training Period

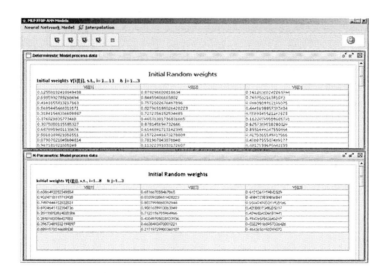

(d) Initial Random Weights Window

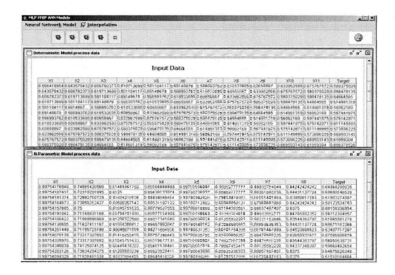

(e) Input Vectors Window and Target Variable

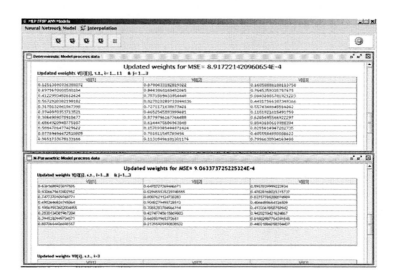

(f) Optimized Weights Window After Training

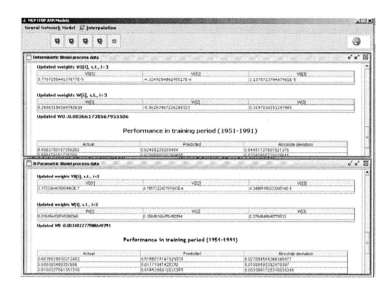

(g) Optimized Weights and Performance Window in Training Period

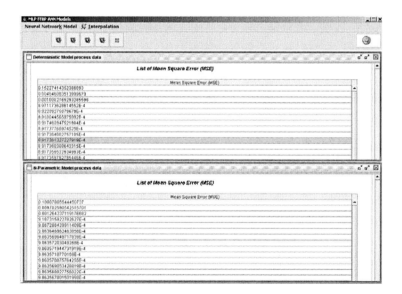

(h) Minimized MSE by Epochs.

Fig. 2.14. Interfaces of Deterministic and n-Parametric BPN Model.

2.14   CONCLUSIONS

There are many types ANNs viz., BPN, CPNN, RBFN, CCNN, PNN, GRNN etc. however in this study, BPN models have been proposed and implemented by using Java.   BPN in deterministic, n-parametric forecast is chosen for three reasons. First, a lot of studies have been made (Section 1.3) on BPN and their abilities, especially in the case of chaos prediction. Second, the availability of meteorological data corresponds to the needs of BPN with input and output data. And third, its use and its construction are easy. Proposed algorithms 2.1 and 2.2 are used for the training of the models. Being a gradient descent algorithm, it minimizes MSE of the models. In the design of the model, selection of number of input vectors and deciding the number of neurons in hidden layer are an extremely difficult tasks and it is a dynamic selection process depending on data time series. It has been identified that using two hidden layers rarely improves the model, and it may introduce a greater risk of converging to a local minima. There is no theoretical reason for using more than two hidden layers and thus single hidden layer has been selected. It has been found that, if the numbers of neurons are increased in the hidden layer, the MAD (% of mean) between actual and model predicted values increases. Thus two or three neurons have been used in the model (*depending on internal dynamics of time series*). It has been identified that MAD (% of mean) is inversely proportional to number of input vectors. Thus eleven or more (depending on internal dynamics of time series) input vectors have been utilized for the model development. In deterministic forecast, meteorological data of n preceding years are used as input to predict the meteorological data of $(n+1)^{\text{th}}$ year. While, in

74

parametric forecast n parameters of ith year are used as input to predict i^{th} year meteorological data as a target.

Chapter 3

BPN MODEL IN DETERMINISTIC FORECAST

BPN model in deterministic forecast, over the smaller region have been developed and verified. The performance of the model have been evaluated and discussed in this chapter.

3.1. INTRODUCTION

It is found that, the ANN has capacity to predict next year data by learning past year data time series called deterministic forecast. In an attempt to predict long-range rainfall over the districts, a BPN model in deterministic forecast has been developed and tested is discussed in this chapter. The performances of model in prediction have been found to be excellent.

3.2 BPN MODEL IN DETERMINISTIC FORECAST

In an attempt to identify pattern and prediction of long-range TMRF over the districts BPN models in deterministic forecast have been developed and verified [45, 46]. The entire development practices including the collection of data, model architecture, training of the model, minimizing the mean square error (MSE), local minima and global minima, performance of the models is offered in the following sections.

3.2.1 Data descriptions and Pre-Processing

In the present study, TMRF time series (1951-2004) have been constructed for Ambikapur districts of Chhattisgarh, India (Geographically the area is located between $23^0 07' 23"$ N, $83^0 11' 39"$ E, elev. 1958 ft, TGA 15733 Sq km). Missing years of data (less than 10%) are replaced by neighboring (within two Kms) raingauge station data. Even after these replacements, monthly rainfall values for few stations and for few years were missing and these were replaced by their monthly mean values. This district of the state has five or more representative

78

raingauge stations. The station averages are calculated only up to the district level. Thus the TMRF data series so constructed are homogeneous spatially as well as temporally. Since training algorithm is limited to the intervals 0 to 1 therefore data normalized by using following formula $R_i = (x_i + \min(x_i)) / (x_i + \max(x_i))$ is used and other hand the equation the equation $x_i = (\min(x_i) - R_i \max(x_i)) / (R_i - 1)$ is used to de-normalize. Data for first 40 years (1951-1991) are used for training the network and data for the remaining period (1992-2004) are used independently for validation.

3.2.2 Architecture

BPN model in deterministic forecast is depicted in Fig. 3.1, where eleven input vectors $(x_1...x_{11})$ in input layer are used to input eleven years' meteorological data time series, three neurons in hidden layer $(z_1...z_3)$ and one neuron (y_k) in output unit are used to observe 12th year TMRF data. Thus a total of 40 $(11 \times 3 + 3 + 3 + 1)$ trainable weights including biases have been used in the network. Output target value for the network is actual 12th year TMRF data. In fact, it has been found that MAD (% of mean) between actual and model predicted value is directly proportional to number of hidden neuron (p). As a result, three neurons in hidden layer have been selected and it provides most desirable result. The neurons output is obtained as $f(x_j)$ where f is a transfer function typically the sigmoid function is used. Learning rate (LR) $\alpha = 0.3$ is used. The framework of the model is shown in Table 3.1. Trainable weights of the network initialized by the random values between 0 and 1 are shown in Fig. 3.1 and in Table 3.2. The selection of parameters as depicted in

79

Fig. 3.1 is an effort. The experimental results on this regard are discussed in the following subsequent section.

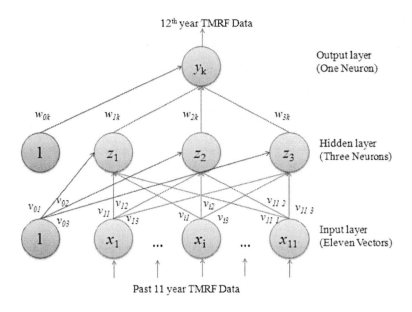

Fig. 3.1. BPN model in deterministic forecast.

Table 3.1. BPN model in deterministic forecast.

Model in	LR (α)	Neuron in Hidden Layer (p)	Input vector (n)	Trainable weights (w_i)	Target (y_k)
Deterministic forecast	0.3	03	Past 11 years TMRF	40 (including biases)	12TH YEAR TMRF

Table 3.2. Initial Set of Weights before training.

Initial weight $V_{ij}; i = 1,2,3,...,11; j = 1,2,3$		
0.5271928310394287	0.049051523208618164	0.1370483636856079
0.44445812702178955	0.04703259468078613	0.44343096017837524
0.9497716426849365	0.3267333507537842	0.4879154562950134
0.8398919105529785	0.191728413105011	0.3600116968154907
0.8808531165122986	0.02883845567703247	0.7311607599258423
0.3852413296699524	0.17276179790496826	0.0242651104927063
0.903407096862793	0.35194820165634155	0.909576952457428
0.40007954835891724	0.35388296842575073	0.8753505945205688
0.6699069142341614	0.8664072155952454	0.035347044467926025
0.93645840883255	0.38544899225234985	0.2427670955657959
0.8964948058128357	0.3399583101272583	0.2699219584465027
Initial weight $V0_{ij}; i = 1,2,3$		
0.7745048999786377	0.23458212614059448	0.545025646686554
Initial weight $W_i; i = 1,2,3$		
0.10550838708877563	0.6436562538146973	0.2813866138458252
Initial W0 : 0.5		

3.2.2.1. Selection of number of neurons in hidden layer (p)

The model is examined with districts and subdivisions TMRF time series [40,44]. As per the result it is concluded that, if number of neurons (p) increases in the hidden layer, the MAD (% of mean) between actual and predicted value is also increases regularly.

Accordingly, following relation between hidden neurons (p) and MAD (% of mean) is obtained-

$$MAD \propto p$$

$$MAD = cp$$

where, c is error constant. The relationship between number of neurons (p) in hidden layer and MAD (% of mean) is shown in Fig. 3.2 (a) and Table 3.3. Relationship between error constant (c) and number of neurons (p) in hidden layer is shown in (b) and Table 2.3. Thus $p = 2$ or 3 (depending on internal dynamics of time series) is sufficient for nearly all forecasting cases we can say.

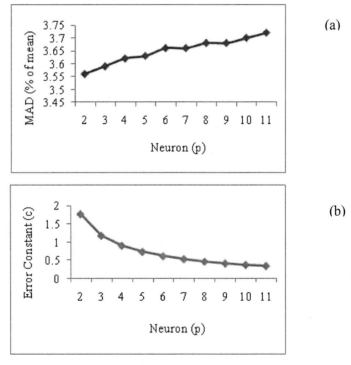

Fig. 3.2. Relation between number of neurons (p) in hidden layer and MAD (% of mean).

Table 3.3. Relation between neurons (p) in hidden layer and MAD (% of mean).

Neuron (p)	MAD (% of MEAN)	Error Constant (c)
2	3.56	1.78
3	3.59	1.18
4	3.62	0.90
5	3.63	0.73
6	3.66	0.61
7	3.66	0.53
8	3.68	0.46
9	3.68	0.41
10	3.7	0.37
11	3.72	0.34

3.2.2.2. Selection of input vectors (*n*)

The relation between number of input (n) and MAD (% of mean) has been shown in Fig. 3.3 and in Table 3.4. on the basis of this experimental result it is observed that MAD (% of mean) is inversely proportional to n. [40,44]. Thus $n = 11$ or more (depending on internal dynamics of time series) input vectors have been selected for this study.

$$MAD \propto \frac{1}{n}$$

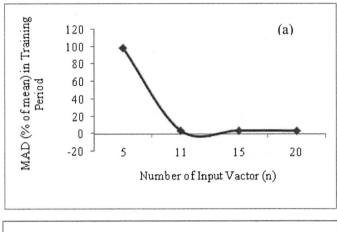

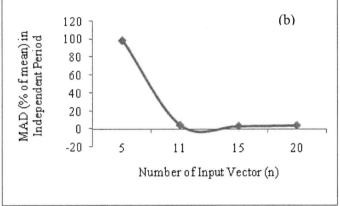

Fig. 3.3. Relation Between Number of Input Vectors (n) and MAD (% of mean) (a) In Independent Period (b) Training Period.

Table 3.4. Relation between Number of Inputs (n) and MAD (% of mean)

Input (n)	MAD (% of mean)		SD (% of mean)	
	Training period	Independent period	Training	Independent
5	98.52	98.54	4.99	4.69
11	3.59	4.03	4.65	5.17

| 15 | 3.51 | 2.96 | 4.75 | 3.88 |
| 20 | 3.27 | 3.69 | 4.28 | 4.69 |

3.2.3   Training of the Model

BPN is depicted in the above Figure 3.1 is trained by using the algorithm 2.1 that is previously discussed in the section 2.7. Wherein, 20, 00000 iterative parallel processes it is generally called 'epochs' are applied to train the network. At this juncture, initial weights before training and optimized weights after training are shown in Table 3.2 and 3.5. The trained network is presented an excellent learning curve. Wherein, an initial MSE that is equal to 8.6719E-04 (local minima) and final MSE very close to the optimum value of 3.8712559948953136E-05 (global minima) have been found as depicted in following Figure 3.4 and Table 3.6.

Table 3.5. Optimized Weights after training

Optimized weight $V_{ij}; i = 1,2,...,11; j = 1,2,3$		
6.270598280340316	-2.0018271973656407	1.6423065173514284
-3.9865007440307214	11.079553010431491	7.553871579443576
0.199126014942202	2.6061153786907916	3.7054459154149
1.8663610907790922	-10.234867498568144	-5.206031833333133
-1.3285773882910707	-1.6445481520677703	-1.4167666469550697
-2.2955962777155356	0.10258439352523269	-0.6287308576104448
-5.060791432018832	-0.5036163178883608	-5.034922160467127

85

6.030420540993752	2.250903868798964	4.040110957750795
-3.6506887651307047	-3.3864106511729157	-3.86532590441335
-3.3102116272578717	2.245673840960485	-1.1181211626486898
8.612823408332893	-3.711303665382227	3.0618059057010174
Optimized weight $V0_i; i = 1,2,3$		
-2.7127385467043E-4	-5.818654669719E-4	2.97760892141918E-4
Optimized weight $W_i; i = 1,2,3$		
6.731181177946909	7.43777627223151	-7.711594995180171
Optimized W0 : -4.6218733807633356E-04		

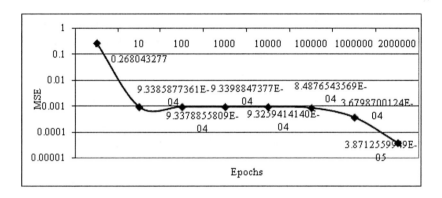

Fig. 3.4. Learning Curve.

Table 3.6. Model Training Results.

Model	Epochs	Minimum MSE (Local Minima)	Final MSE (Global Minima)
BPN	20,00000	8.67198916540E-04	3.871255994895E-05

3.2.4 Performance of the Model (Results Analysis)

BPN model in deterministic forecast has been tested. The performance of the model during the training period and independent/testing period is given in the Table 3.7 and 3.8 respectively. Wherein, the actual TMRF, predicted TMRF time series, and absolute deviation between them in the training as well as independent/testing period are presented. Following significant results have been found.

1.   It is found that, the MAD (% of mean), i.e., 3.3, is exceptionally less than the SD (% of mean) i.e., 21.6, during the training period (1951-1991). And The CC between actual and predicted TMRF (in mm.) was obtained 0.98.   It implies that, the model was successfully trained with 20, 00000 epochs at the level of MSE 3.8712559948953136E-05 (global minima).

2.   Thus, the model performs extremely well in the training period.

3.   However, the performance of the model in the independent period (1992-2004) shows very unfortunate. Because, MAD (% of mean), i.e., 312.0 mm. is roughly equal to the SD (% of mean) i.e., 353.8 mm. during the training period. And CC between actual and predicted TMRF (in mm.) is -0.2.

4.   In these facts, it can be concluded that, the deterministic ANN model although found excellent in the training period however unfortunate in the independent/testing period.

5.   But, by carefully considering Figure 3.5 to 3.8 and Table 3.7 to 3.9. It is cleared that the model is extremely ineffective only for the year 1994 and 2004. For other years the model has produced good result.

6.   The average TMRF of this region from 1951 to 2004 is 1242.7 mm. TMRF for the year 1994 is 2092.8 mm. which is much unexpected. Similarly, TMRF for the year 2004 is 858.4 mm. which is also unexpected minimal TMRF over this region during the independent period. In both the years TMRF shows high variation from the average TMRF 1242.7 mm.

7.   If we replace actual TMRFs of these years from the average TMRF (i.e., 1293 mm.) then it is observed that, the CC between actual and predicted value is near about 0.6. The MAD (% of mean) is 15.1 that is less than the SD (% of mean) is 20.2. Which indicated model is sensible for TMRF prediction by past recorded data time series. This is the justification of usefulness of the model.

8.   And as a final point, it is concluded that, this model in deterministic forecast may be utilized for all standard cases.

9.   BPN model may able to predict future data by its past recorded time series.

10. Finally, it is concluded that BPN model in deterministic forecast may able to identify internal dynamics of chaotic motion and predict it.

Table. 3.7. Performance of the BPN model in training period (1951-1991).

Year	1	2	3	Actual TMRF (in mm.)	Predicted TMRF (in mm.)	Absolute Deviation (in mm.)
1962	0.58050	0.57603	0.00447	952.3	920.2	32.1
1963	0.59860	0.60157	0.00296	1089.6	1113.3	23.7
1964	0.64675	0.64404	0.00270	1523.4	1495.9	27.5
1965	0.61510	0.61907	0.00396	1226	1260.6	34.6
1966	0.56075	0.55996	7.96E-4	815.4	810.1	5.3
1967	0.59756	0.60316	0.00559	1081.4	1126.2	44.8
1968	0.57527	0.58371	0.00844	914.8	975.8	61.0
1969	0.61127	0.61321	0.00193	1193.3	1209.8	16.5
1970	0.57402	0.58042	0.00639	906	951.7	45.7
1971	0.66960	0.66908	5.19E-4	1773.5	1767.4	6.1
1972	0.61071	0.61449	0.00378	1188.6	1220.8	32.2
1973	0.60649	0.61261	0.00611	1153.4	1204.7	51.3
1974	0.57592	0.57718	0.00126	919.4	928.4	9.0
1975	0.64784	0.64744	4.03E-4	1534.6	1530.5	4.1
1980	0.66087	0.66281	0.00193	1674	1695.6	21.6
1981	0.60033	0.60494	0.00460	1103.4	1140.7	37.3
1983	0.57837	0.57848	1.08E-4	936.9	937.7	0.8
1984	0.60077	0.60272	0.00195	1106.9	1122.6	15.7
1985	0.62385	0.61282	0.01102	1303.2	1206.5	96.7
1986	0.61841	0.62062	0.00221	1254.8	1274.3	19.5
1987	0.61021	0.59413	0.01607	1184.4	1054.6	129.8
1988	0.61576	0.60766	0.00809	1231.7	1163.1	68.6
1989	0.61875	0.61679	0.00196	1257.8	1240.7	17.1
1990	0.64837	0.63635	0.01201	1540	1419.9	120.1
1991	0.65830	0.66522	0.00691	1645.6	1722.9	77.3

Table.3.8. Performance of the BPN model in independent/testing period
(1992-2004).

Year	1	2	3	Actual TMRF (in mm.)	Predicted TMRF (in mm.)	Absolute Deviation (in mm.)
1992	0.61096	0.61537	0.00441	1190.7	1228.4	37.7
1993	0.62075	0.64769	0.02693	1275.5	1533.0	257.5
1994	0.69481	0.59283	0.10197	2092.8	1044.5	1048.3
1995	0.60567	0.59951	0.00616	1146.7	1096.8	49.9
1998	0.59345	0.59855	5.09E-03	1049.3	1089.2	39.9
1999	0.61550	0.60566	0.04983	1229.5	1146.6	82.9
2000	0.61625	0.59366	0.02259	1236	1050.9	185.1
2001	0.67357	0.62608	0.06748	1820.5	1323.5	497.0
2002	0.59815	0.55955	0.03859	1086	807.5	278.5
2003	0.61678	0.58812	2.87E-02	1240.6	1008.6	232.0
2004	0.56715	0.65227	0.08511	858.4	1580.8	722.4

1-Actual Rainfall (Normalized)

2-Predicted TMRF (Normalized)

3-Absolute Deviation (Normalized)

Table 3.9. Performance of BPN in deterministic forecast during training
(1951-2004) and independent Period (1992-2004).

Data	Training period (1951-1991)					
	Mean	SD	MAD	SD (% of mean)	MAD (% of mean)	CC
Normalized	0.61	0.034	0.005	4.78	0.76	0.98
De-Normalized (in mm)	1220.4	263.3	39.9	21.6	3.3	0.98

Data	Independent/testing period (1991-2004)					
	Mean	SD	MAD	SD (% of mean)	MAD (% of mean)	CC
Normalized	0.62	0.034	0.04	5.50	6.41	0.5
De-Normalized (in mm)	1293.3	353.8	312.0	27.4	24.1	0.5

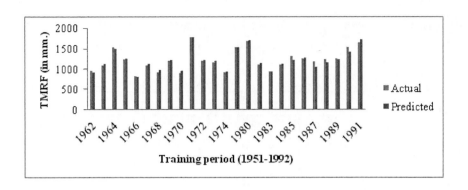

Fig. 3.5. Performance of BPN in deterministic forecast during the training
period (1951-1991).

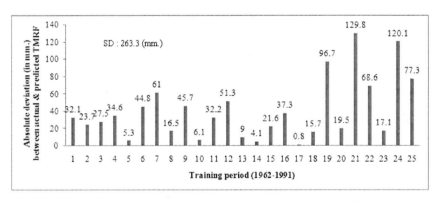

Fig. 3.6. Absolute deviation between actual TMRF and predicted TMFR
during training period.

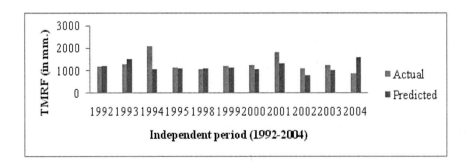

Fig. 3.7. Performance of BPN in deterministic forecast during the
independent/testing period (1992-2004).

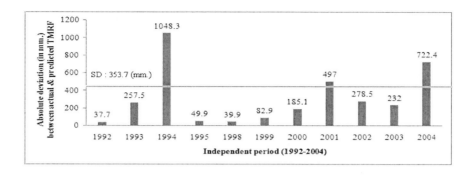

Fig. 3.8. Absolute deviation between actual TMRF and predicted TMFR
during independent /testing period.

3.5   CONCLUSIONS

In the existing work, the BPN model in deterministic forecast
developed and verified to predict long-range monsoon rainfall over very
smaller geographical region of India. The performance of the model is
depicted that, MAD (% of mean) is very less than from the SD (% of
mean) in the training period. The CC between actual rainfall and model
predicted rainfall value is more than 0.98. This observation clearly
indicates the effectiveness of the model in prediction during training
period. However, this model shows poor performance due to unexpected
TMRF for the year of 1994 and 2004. The model in deterministic forecast
unfortunately fails to forecast TMRF for these years from past recorded
TMRF data time series. For other year's model performed noteworthy.
Thus it may be concluded that the ANN model performs healthy to
recognize internal dynamics of TMRF which generally representing

93

chaotic in nature and very difficult to forecast and prediction as compared to the statistical model as we discussed in the chapter 1. Moreover, forecasts have a longer lead-time as they can be made a year in advance. The model developed in this chapter used to forecast districts long-range TMRF can forecast noteworthy. One can consider another application of ANN, i.e., ANN model in parametric forecast to predict TMRF over the district. In this application some parameters those are physically linked with the TMRF over that region as input to observed TMRF of the region as an output. However, identification of its dynamic parameters by observing their physical linkage and degree of relationship with TMRF is extremely difficult. The identification of these parameters and application of ANN in parametric forecast is presented in the next chapter. Finally it is concluded that the ANN model in deterministic forecast is capable to explain the non-linearity of the TMRF time series data. Here, developed Java based simulator for this model is given in the following Figure 3.9. Output screens are shown here for 2000000 epochs.

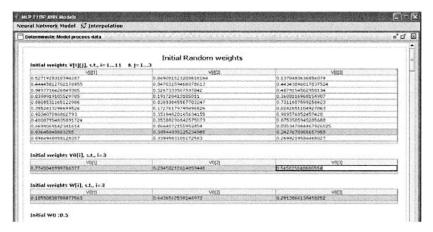

(a)

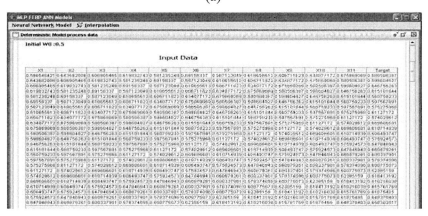

(b)

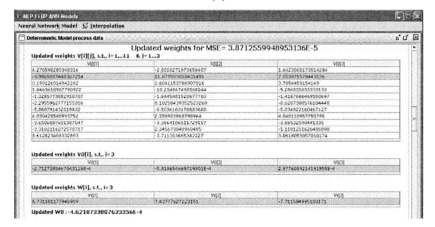

(c)

95

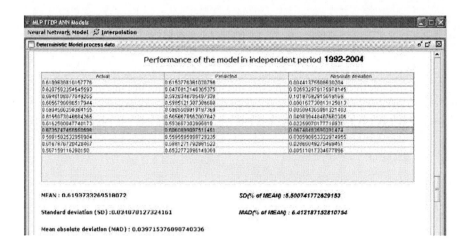

(d)

Fig. 3.9. Java based BPN model output screens in deterministic forecast.

Chapter 4

BPN MODEL IN PARAMETRIC FORECAST

In this chapter, the development of BPN in n-parametric forecast model is explained. An 8 parameters BPN model has been developed for long-range monsoon rainfall pattern recognition and prediction for the Ambikapur region. This model is compared with deterministic BPN model is also discussed.

4.1. INTRODUCTION

To predict long-range TMRF (in mm.) over the smaller scale geographical region "district", eight parameters BPN model has been developed. The collection of data, identification of dynamic parameters those are physically connected with the TMRF of Ambikapur district region, processing of the data, model skeleton, training of the network, minimizing MSE during the training process, performances of the model and their evaluation over the deterministic BPN model have been discusses in the following sections.

4.2. COLLECTION AND PREPROCESSING OF DATA

Ambikapur region of Surguja district (IMD station index No. 42693) of Chhattisgarh has been considered for the study as shown in Figure 1.2 (chapter 1). Geographically the area is located between $21^0 43'$ to $24^0 12'$ N lat., and $81^0 01'$ to $83^0 51'$ E long (TGA 15733 Sq km).

Table 4.1. Identified 8 input parameters those are physically connected to TMRF.

No.	Parameters	Abb.	Month	Correlation coefficient (Period: 1981-2004)
P1	Mean Wet Bulb Temperature (in Deg. C)	WBT	Jun. (Prev. year 12 HR)	0.41

P2	Mean Dew Point Temperature (in Deg. C)	DPT	Jun. (Prev. year 03 HR)	0.45
P3	Mean Relative Humidity (in %)	RH	Aug. (Prev. year 03 HR)	-0.45
P4	Mean Vapour Pressure (in hPa)	VP	Jun. (Prev. year 03 HR)	0.55
P5	Mean Total Cloud Amount (in Oktas)	TOC	Oct. (Prev. year 03 HR)	-0.44
P6	Mean Station Level Pressure (in hPa)	SLP	Feb. (Current year 03 HR)	-0.41
P7	Highest Maximum Temperature (in Deg. C)	HMAX	Jan. (Current year)	0.56
P8	Previous year rainfall (in mm.)	PYRF	Jun. (Prev. year)	-0.42

IMD raingauge stations of this region having maximum meteorological data (1951-2004) have been selected. Eight regional meteorological parameters are identified as input. The detail of eight parameters used in this model has been presented in Table 4.1. All these eight parameters are chosen after careful investigation by seeing their physical linkage and degree of relationship with the TMRF. The CCs of each parameter with TMRF for the 24 years period 1981-2004 have also been provided in Table 4.1.

4.3.   BPN MODEL PARAMETRIC FORECAST

Eight parameters BPN model is depicted in the following Fig. 4.1. Wherein, eight input parameters which are physically connected with TMRF over this region are used to input in the BPN. Three neurons in hidden layer, one neuron in output layer, a total of 8.3+3=27 trainable weights including 3 biases in hidden layer and 1 bias in output layer have been used to train the model and observe TMRF as model output. It is observed that three neurons can be utilize in hidden layer therefore three neurons in hidden layer is chosen. The detail discussion of selection of neurons in hidden layer is given in following subsection. The neuron output is obtained as $f(x_j)$. Where f is a transfer function typically the sigmoid function.   The skeleton of the model is shown in Fig. 4.1 and Table 4.2.

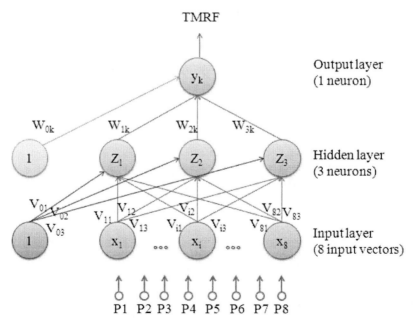

Fig. 4.1. 8-Parameters BPN Model.

Table 4.2. 8 Parameters BPN Model.

Model input	Model type	LR	Neuron in Hidden Layer (p)	Input vector (n)	Trainable weights (w_i)	Target (y_k)
8 parameters (P1,P2,...,P8)	BPN in parametric forecast	0.3	03	08	27 including biases	TMRF

4.3.1. Selection of neurons in hidden layer

An experiment is done in which with the random initial weights including biases initially selected randomly, and with different value of 'p' i.e., 1 to 22, the neural network is trained by using 100 epochs. And each selection 'p' the MAD (% of mean) between actual and predicted value is obtained in training and independent period are given in Table 4.3 and depicted in the Fig. 4.2 and 4.3. On the basis of the experiment, three neurons in hidden layer is selected. It is clear that, MAD (% of mean) is directly proportional to the number of neurons in hidden layer (p). It indicates while 'p' increases, the MAD (% of mean) is gradually increases. Testing started with single neuron up to 22 neurons. It is found that 1 neuron in hidden layer produces minimum MAD (% of mean) during the training as well as independent period that is 4.039053286 and 3.666168517 respectively. However, in this study 3 neurons in hidden layer are used instead of single neurons. Two basic reason of this has been observed-

1. MAD (% of mean) i.e., 3.650164938 in the independent period (when 2 neurons are used) is less than the MAD (% of mean) i.e., 3.666168517 (when 1 neuron is used) given in the Table 4.3.

2. However, three neurons will provide more trainable (i.e., adjustment) weights instead of one neuron or two neurons.

4.3.1.1. System Break-Down

It is found that, MAD (% of mean) is directly proportional to the number of neurons in the hidden layer (p). The experiment started with single neurons up to 22 neurons in the hidden layer. Through each increment of 'p', the MAD (% of mean) between actual and predicted values is given in the Table 4.3. It is observed that, with increment of 'p' the MAD (% of mean) is gradually increases in both training and independent period (see Fig. 4.4 a & b) up to 21st neuron. When 22nd neuron is introduced the MAD (% of mean) is unexpectedly increased in the both the period that is 62.47148378 and 61.38902481 respectively (see Fig. 4.5. a & b). While SD (% of mean) of actual TMRF (in mm.) in the training and independent period is 5.114626053 and 5.500741773 respectively. These are exceptionally less than the MAD (% of mean). This observation indicates that the model with 22 neurons in hidden layer is ineffective in prediction of TMRF. We can say 22nd neuron is the system breaking neuron and 22 is system break-down point. In general, it is concluded that, for all forecasting applications such break-down point will come into view. Therefore we cannot place any number of hidden neurons in the hidden layer. We have to identify break-down point first and then may place number of neurons in the hidden layer (p) which is less than this point. It is noted that, 2 or 3 neurons in hidden layer is noteworthy for all forecasting applications.

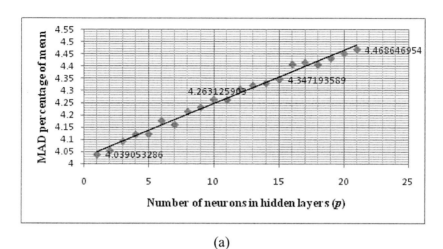

(a)

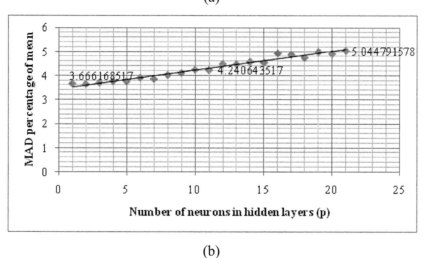

(b)

Fig. 4.2. Relationship between MAD (% of mean) and number of neurons in hidden layer. (a) Training period (b). Independent period.

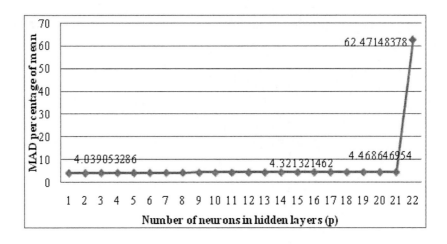

(a)

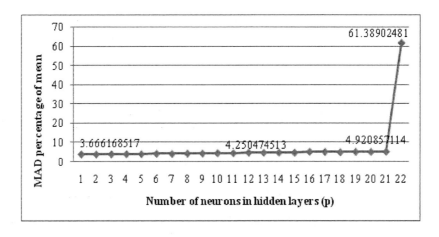

(b)

Fig. 4.3. System breaking point. (a) Training period (b). Independent period.

Table 4.3. Neurons in hidden layer.

p	MAD (% of mean) in the Training period (1951-1991)	MAD (% of mean) in the independent period (1992-2004)
1	4.039053286	3.666168517
2	4.055213523	3.650164938
3	4.092718325	3.710746478
4	4.120621746	3.767731212
5	4.122166012	3.763576406
6	4.176529918	3.915572068
7	4.160463518	3.862557297
8	4.214445559	4.030392003
9	4.231021855	4.108467303
10	4.263125903	4.250474513
11	4.262055966	4.240643517
12	4.306004583	4.490832652
13	4.321321462	4.477708019
14	4.331131529	4.576958221
15	4.347193589	4.547129395
16	4.407802525	4.937666834
17	4.415292465	4.881408884
18	4.406974846	4.755268358
19	4.432885875	4.994476746
20	4.452409052	4.920857114
21	4.468646954	5.044791578
22	62.47148378	61.38902481

4.3.2. Training of the Model

Algorithm 2.2 (Section 2.7) is used to train the network. The initial random trainable weights including biases on hidden and output layer as shown in Table 4.3 is selected randomly by using "Random" class of Java is optimized by exercising 20, 00000 epochs. The optimized weights are shown in Table 4.4. The trained network presented excellent learning curve started with local minima MSE = 0.0010583407253476 to global minima MSE = 6.555600546440187E-04 that is exceptionally close to the optimum value of 0 as shown in the Fig. 4.4.

Table 4.4. Initial set of weights before training.

Initial weights $V_{ij}; i = 1,2,3,...8; j = 1,2,3$		
0.861691415309906	0.22715413570404053	0.8768165707588196
0.7864313721656799	0.1989791989326477	0.3333062529563904
0.3668724298477173	0.438451886177063	0.01480400562286377
0.07771182060241699	0.2873939871788025	0.560234010219574
0.3179476857185364	0.6397307515144348	0.6804287433624268
0.5102882385253906	0.7174369692802429	0.44445663690567017
0.3175405263900757	0.23509055376052856	0.25821787118911743
0.6618416905403137	0.286279559135437	0.4285861849784851
0.861691415309906	0.22715413570404053	0.8768165707588196
Initial weights $V0_{ij}; i = 1,2,3$		
0.2999613285064697	0.2589431405067444	0.2390645146369934
Initial weights $W_i; i = 1,2,3$		
0.10159945487976074	0.19197046756744385	0.545210063457489
Initial W0 : 0.5		

106

Table 4.5. Optimized set of weights before training, 20,00000 epochs.

Optimized weights V_{ij}; $i = 1,2,3,...8$; $j = 1,2,3$		
1.1504005495726712	-3.231188742943418	-4.49444937122248
8.145844471279098	-9.468212927557897	1.626150240201612
0.7304107553183267	-3.5220397850993557	-9.165839546716056
1.2818430196861041	-1.653726284147089	18.101521791435896
6.2718139989280415	10.830371711169267	-1.3520536437384232
-12.26532261361735	13.31525241952504	6.86774036208046
-2.7422444068742533	4.899050852156461	-8.019181173086945
0.17076326117063892	-2.629295308689212	-8.920492820789686
Optimized weights $V0_{ij}$; $i = 1,2,3$		
-3.79105590770616E-4	2.75069112419501E-4	5.949836614418309E-4
Optimized weights W_i; $i = 1 to 3$		
-2.6378943872281324	2.2084526467128085	7.2276776719864335
W0 : 8.011408367628631E-04		

Table 4.6. Training Results of Parametric ANN Model.

Target	Epochs	Minimum MSE (Local Minima)	Final MSE (Global Minima)
TMRF	20,00000	0.0010583407253476	6.555600546440187E-04

107

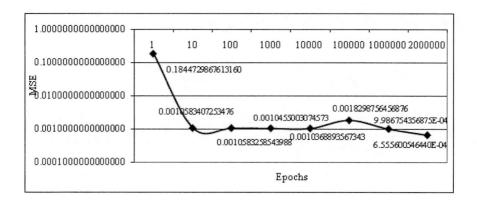

Fig. 4.4. Minimizing error (MSE) during training process

4.2.4 Performance of the Model

BPN model in parametric forecast has been developed and evaluated. The performance of the model during the training period and independent/testing period is given in the Table 4.7 and 4.8 respectively. Wherein, the actual TMRF, predicted TMRF, and absolute deviation between them in the training as well as independent/testing period are presented. It is found that MAD (% of mean) is incredibly less than the SD (% of mean) during the training period and also during the independent period. Correlation between actual and model predicted rainfall values is equal to 0.8 as shown in Table 4.9. This result is clearly indicated that the effectiveness of the model in accuracy of prediction. Carefully look at, Fig. 4.5 and 4.6. The bars are representing difference between actual and predicted values in the training period. Absolute deviation (see Fig. 4.6) shows lesser than the standard deviation (i.e., 296.3 mm.) almost all cases except the year 1952, 1956, 1957, 1975, 1980. In other hand in independent period (see Fig. 4.8) the absolute deviation is less than for all cases except the year, 1994, and 2001 only.

108

BPN model in parametric forecast also better evaluated over the deterministic forecast. It has been already observed that, the deterministic model could not explain the TMRF in the year of 1994 and 2004. The parametric model is slightly superior explaining the variability of these years.

Table. 4.7. Performance of the BPN model in training period (1951-1991)

Year	1	2	3	Actual TMRF (in mm.)	Predicted TMRF (in mm.)	Absolute Deviation (in mm)
1952	0.64368	0.60677	0.03690	1492.2	1155.7	336.5
1953	0.60690	0.61739	0.01049	1156.8	1245.9	89.1
1954	0.61983	0.60246	0.01736	1267.3	1120.5	146.8
1955	0.58123	0.61181	0.03057	957.6	1197.9	240.3
1956	0.69158	0.62984	0.06173	2049	1358.2	690.8
1957	0.58712	0.63021	0.04308	1001.1	1361.6	360.5
1958	0.61065	0.61675	0.00610	1188.1	1240.3	52.2
1959	0.60671	0.61920	0.01248	1155.2	1261.7	106.5
1960	0.63407	0.60920	0.02487	1398.1	1175.9	222.2
1961	0.67586	0.66101	0.01484	1848.2	1675.6	172.6
1962	0.58050	0.60874	0.02823	952.3	1172.0	219.7
1963	0.59860	0.62288	0.02427	1089.6	1294.5	204.9
1964	0.64675	0.64058	0.00617	1523.4	1461.3	62.1
1965	0.61510	0.61060	0.00450	1226	1187.6	38.4
1966	0.56075	0.59234	0.03158	815.4	1040.8	225.4
1967	0.59756	0.61269	0.01513	1081.4	1205.4	124.0
1968	0.57527	0.56582	0.00944	914.8	849.4	65.4
1969	0.61127	0.60450	0.00677	1193.3	1137.0	56.3

1970	0.57402	0.58908	0.01505	906	1015.9	109.9
1971	0.66960	0.66796	0.00163	1773.5	1754.4	19.1
1972	0.61071	0.61593	0.00521	1188.6	1233.2	44.6
1973	0.60649	0.62579	0.01930	1153.4	1320.9	167.5
1974	0.57592	0.61017	0.03425	919.4	1184.1	264.7
1975	0.64784	0.59149	0.05635	1534.6	1034.2	500.4
1980	0.66087	0.61235	0.04852	1674	1202.5	471.5
1981	0.60033	0.61123	0.01089	1103.4	1193.0	89.6
1983	0.57837	0.61535	0.03698	936.9	1228.2	291.3
1984	0.60077	0.60773	0.00696	1106.9	1163.7	56.8
1985	0.62385	0.62822	0.00437	1303.2	1343.1	39.9
1986	0.61841	0.61746	9.50E-4	1254.8	1246.5	8.3
1987	0.61021	0.59598	0.01423	1184.4	1069.0	115.4
1988	0.61576	0.62718	0.01142	1231.7	1333.6	101.9
1989	0.61875	0.61373	0.00502	1257.8	1214.2	43.6
1990	0.64837	0.62209	0.02627	1540	1287.4	252.6
1991	0.65830	0.64661	0.01168	1645.6	1522.0	123.6

1- Actual Rainfall (Normalized)
2- Predicted TMRF (Normalized)
3- Absolute Deviation (Normalized)

Table.4.8. Performance of the parametric BPN model in
independent/testing period (1992-2004)

Year	1	2	3	Actual TMRF (in mm.)	Predicted TMRF (in mm.)	Absolute Deviation (in mm.)
1992	0.61096	0.64197	0.03101	1190.7	1475.1	284.4
1993	0.62075	0.61540	0.00535	1275.5	1228.6	46.9
1994	0.69481	0.73365	0.03884	2092.8	2703.2	610.4
1995	0.60567	0.61081	0.00513	1146.7	1189.4	42.7
1998	0.59345	0.59027	0.00318	1049.3	1024.9	24.4
1999	0.61550	0.61139	0.00411	1229.5	1194.3	35.2
2000	0.61625	0.61686	6.09E-4	1236	1241.3	5.3
2001	0.67357	0.61703	0.05653	1820.5	1242.8	577.7
2002	0.59815	0.61074	0.01259	1086	1188.9	102.9
2003	0.61678	0.60267	0.01411	1240.6	1122.2	118.4

1- Actual Rainfall (Normalized)
2- Predicted TMRF (Normalized)
3- Absolute Deviation (Normalized)

Table 4.9. Performance of parametric BPN in during training and
independent Period

Data	Training period (1951-1991)					
	Mean	SD	MAD	SD (% of mean)	MAD (% of mean)	CC
Normalized	0.62	0.03	0.02	5.51	3.21	0.61
De-Normalized (in mm)	1257.8	293.6	174.7	23.3	13.8	0.61

111

Data	Independent/testing period (1991-2004)					
	Mean	SD	MAD	SD (% of mean)	MAD (% of mean)	CC
Normalized	0.62	0.03	0.02	5.50	2.96	0.8
De-Normalized (in mm)	1293.2	353.7	188.2	27.4	14.5	0.8

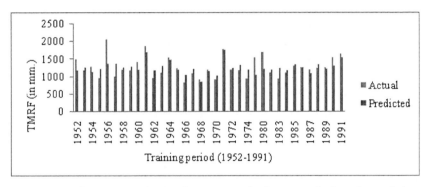

Fig. 4.5. Performance of BPN in parametric forecast during the training period (1951-1991)

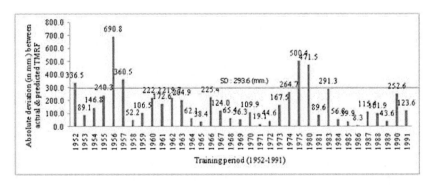

Fig. 4.6. Absolute deviation between actual TMRF and predicted TMRF during training period (1951-1991)

112

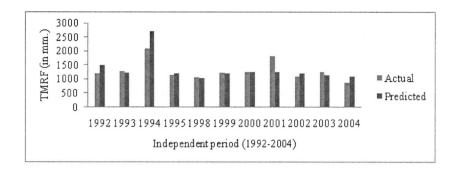

Fig. 4.7. Performance of BPN in parametric forecast during the
independent/testing period (1992-2004)

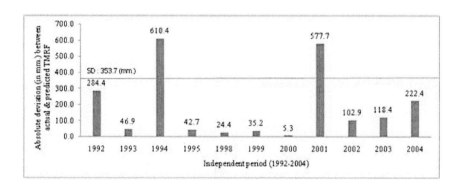

Fig. 4.8. Absolute deviation between actual TMRF and predicted TMFR
during independent period/testing period.

4.5   CONCLUSIONS

BPN model in parametric forecast is developed for prediction of
long-range TMRF over the smaller region and assessed in the training as
well as independent period. The entire development process is discussed
in this chapter. It is observed that the BPN model in parametric forecast
can perform well for pattern recognition as well as prediction however

113

the performance is entirely depending on selection of input parameters. In this study, it is found that identification of input parameter those are physically connected with the TMRF is a huge effort. However have found 8 significant parameters through which model developed and evaluated. It is undoubtedly concluded that this model is better evaluated over the deterministic BPN model (Chapter 3) in this case. It is also concluded that the BPN model in parametric forecast is extremely effective for long-range TMRF over the smaller as well. However it is matter of the selection of predictors. More significant parameter (if we able to search) may produce better result. However, one problem may associate with this phenomenon. That is more input parameters will introduce more unknown variable (i.e., trainable weights) in the network to be trained. To solve this problem we may go for principal componenet analysis to reduce parameters and thed principal components may use as input parameters. In the next chapter we introduce this concept white here, developed Java based simulator for this model is given in the following Figure 4.9. Output screens are shown here for 2000000 epochs.

(a)

114

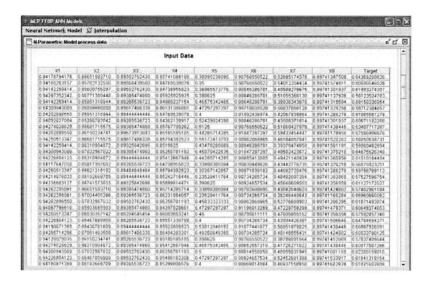

(b)

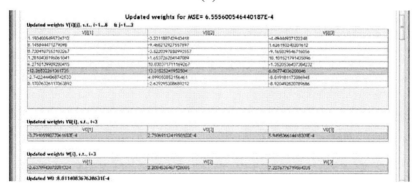

(c)

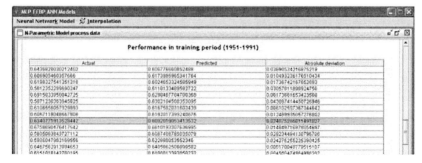

(d)

115

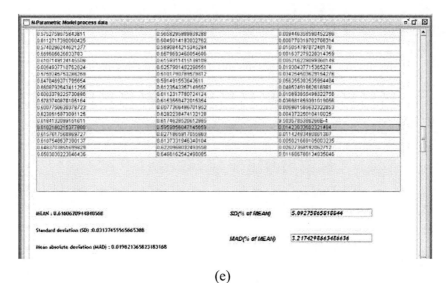

(e)

(d)

Fig. 4.9. Java based BPN model output screens in parametric forecast.

Chapter 5

PRINCIPLE COMPONENTS BPN MODEL

Principal component analysis (PCA) is accomplished over eight parameters (discussed in previous chapter 4) and obtained principal components (PCs) planned to reduce input parameters as well as unknown trainable weights. And a separate BPN model is developed for prediction of TMRF over the same region wherein these PCs are inputted to observe TMRF as target. The performances of the model is explained by strong relationship between dependent (i.e., TMRF), and independent variables (i.e., PCs).

117

5.1.   INTRODUCTION

To predict long-range TMRF (in mm.) over the Indian smaller region "District", BPN model in deterministic forecast as well as parametric forecast (chapter 3 and 4) have been developed and verified. It is observed that model in parametric forecast is better evaluated over the deterministic forecast. However, in parametric model eight parameters have been inputted to observe TMRF as output. Total 31 trainable weights including biases are used in the network. These are unknown weights in the network to be trained. Due to that, it introduced more time and space complexity during the training process. It may affect the network performance as well. To overcome this shortcoming, one can assume to reduce principal components (PCs) of these parameters instead of eight by applying principle component analysis (PCA). In this chapter, PCA over the identified seven parameters (chapter 3) is accomplished and obtained two PCs and used as input vector in BPN model. It is found that, two neurons are suitable in hidden layer. Therefore, 31 trainable weights are reduced to 09 trainable in this model to observe performance. Model is trained successfully by 20, 00000 epochs and verified independently and moreover subsequently model is evaluated over the deterministic as well parametric model is presented in this chapter.

5.2.   PCA OF INPUT VARIABLES

Most important PCA data reduction technique is applied where the objective is to aggregate or merge the information contained in the seven data sets (Table 5.1) into smaller information pieces. Often found those climate variables under the study are highly correlated and such

118

they are effectively saying the same internal dynamics [140]. For example, CC between DPT and VP is 0.98 (Table 5.1). It may be useful to transform the original set of variables to a new set of uncorrelated variable called PCs. These PCs are linear combinations of original variable and are derived in decreasing order of importance so that the first PC accounts for as much possible of the variation in the original data [142]. In other hand, as we know that more input vector will introduce unknown variables (i.e., weights) to be trained and will increase training time as well. Accordingly, a PCA over these parameters have been prepared and described in the succeeding paragraph.

Let x_1, x_2, x_3,..., x_7 are the climate variables are under the study as given above Table 5.1 then the first PC may be defined as

$$Y_1 = a_{11}x_1 + a_{12}x_2 + a_{13}x_3 + ... + a_{18}x_8$$

Table 5.1. Original parameters

YEAR	DPT (x_1)	RH (x_2)	VP (x_3)	TOC (x_4)	HMAX (x_5)	PYRF (x_6)	SLP (x_7)
1952	21.2	89	25.3	1.8	27.2	531.4	946.4
1953	18.5	90	21.9	1.5	27.2	567.9	947.7
1954	18	89	21.8	1.6	27.7	271.7	945.7
1955	21.6	87	26	1.9	27.7	488.7	944.5
1956	18.7	85	22.6	2.8	27.7	259	945.8
1957	22	91	26.8	2.9	26.9	653.7	945.9
1958	18.7	88	21.9	2	29.3	297.6	945.6
1959	17.1	85	20.6	3.7	27.7	324.1	945.7

119

1960	21.3	87	25.4	3.5	27.2	509.3	946.7
1961	19.6	92	23.3	2.5	29	743.1	944.8
1962	21.3	91	25.5	4.2	27.8	377.2	947.3
1963	20.1	86	23.2	2.4	27.7	251.3	947.6
1964	22.5	87	27.3	2.6	29	271.6	946.2
1965	20.1	88	24.1	2.5	26.7	448.9	945.4
1966	18.6	85	21.8	1.4	27.6	322.9	946.2
1967	21.1	73	25.5	1.1	28.5	335.1	945.6
1968	19.7	88	23.5	0.6	27	427.4	946.3
1969	21.5	86	25.9	1.9	28.2	366	946.1
1970	21.2	87	25.2	1.4	27.3	315.4	946.6
1971	22.4	85	27.2	0.6	27	233.2	944.7
1972	22.5	89	27.3	2.7	27.8	533.6	945
1973	17.7	87	20.4	2.9	30	399.4	947
1974	21.8	84	26.3	3	27.4	407.6	946.1
1975	20.6	85	24.6	2	27	333.4	947.2
1980	20.5	88	24.5	3.8	29.2	483.6	946.7
1981	22.6	91	27.9	3.2	27	419.6	946.6
1983	19.6	85	23.3	1.9	27.2	242	946.8
1984	20.1	87	24.1	2.8	26.5	291.5	946.7
1985	22.5	89	27.3	3.3	27.5	278	943.5
1986	20.6	89	24.4	2.9	28.2	524.9	947.4
1987	19.7	85	23.7	2	26.6	393.1	948.2
1988	20.5	85	24.4	2.4	28.2	251.9	946.1
1989	21.3	88	25.5	2.3	29	433.9	946.3
1990	21.7	84	26.1	1.3	30.2	287.2	946
1991	22.3	83	26.9	3.9	28.6	271.2	947.3
1992	21.5	86	26.2	2.5	29.2	642.4	946
1993	19.9	86	23.7	1.6	28.3	537.3	946.7
1994	22	87	26.6	1.1	29.2	140.2	945.6

1995	22.1	89	26.9	2.8	25.4	286.7	945.6
1998	21.4	89	25.9	2.1	24.9	480	947.5
1999	20.9	86	25	4.2	28.2	357.5	948
2000	21.4	92	25.9	3.3	28.2	352.4	945.1
2001	23	83	28	1.2	28.2	297.2	944.8
2002	22.4	88	25.8	2.5	27.9	317.8	948.4
2003	21.8	88	26.2	1.7	28.3	471.1	946.6

such that variance of Y_1 is as large as possible subject to the condition that

$$a_{11}^2 + a_{12}^2 + a_{13}^2 + \ldots + a_{18}^2 = 1$$

this constraints is introduced because if this is note done, then var (Y_1) can be increased simply by multiplying any a_{1j}s by a constant factor. The second PC is defined as

$$Y_2 = a_{21}x_1 + a_{22}x_2 + a_{23}x_3 + \ldots + a_{28}x_8$$

such that var (Y_2) is large as possible next to var $(Y1)$ subject to the constraints

$$a_{21}^2 + a_{22}^2 + a_{23}^2 + \ldots + a_{28}^2 = 1 \text{ and}$$

$$\text{cov } (Y_1, Y_2) = 0 \text{ and so on.}$$

It is quite possible the first two PCs account for most of the variability in the original data. If so then two PCs can be replace the initial seven variables. Consider the variables DPT (x_1), RH (x_2),VP (x_3),TOC (x_4),HMAX (x_5),PYRF (x_6), and SLP (x_7) are on the study as shown in Table 5.1. Find the Tables from 5.2 to 5.7. If we want to reduce the number of variables, we can use the Kaiser criterion [143-145]. Wherein, we will consider only the eigenvalues that they are larger than their average. The average of the eigenvalues is 2294.1 (Table 5.4). Wherein, variable Y_1 is found whose eigenvalue is 16041.4. This is undoubtedly more than the average of the eigenvalue. As a result entire seven variables can be reduced into a single variable Y_1 (Table 5.6). Also see Table 5.6, it is found that $Y_1 >= 0$ and $Y_2 >= 0$ is positive in the number 1 2 3 4 5 6 7 8 9 10 11 12 13 14 15 16 17 18 19 20 21 22 23 24 25 26 27 28 29 30 31 32 33 34 35 36 37 38 39 40 41 42 43 44 45. However, variable Y_2 cannot considered for the study, for the reason that its eigenvalue is 8.3 which is less than the average of the eigenvalues i.e., 2294.1.

Table 5.2. Covariance Matrix for the first seven variables

VA\VA	DPT	RH	VP	TOC	HMAX	PYRF	SLP
DPT	2	0	3	0	0	-3	0
RH	0	10	0	1	-1	152	0
VP	3	0	4	0	0	-2	0
TOC	0	1	0	1	0	14	0
HMAX	0	-1	0	0	1	-2	0
PYRF	-3	152	-2	14	-2	16040	5
SLP	0	0	0	0	0	5	1

Table 5.3. Variables and variances

Variable	Average	Standard Dev	Variance(Var)
PYRF	386.8848	126.6488	16039.9262
RH	87	3.1056	9.6444
VP	24.9304	1.916	3.6711
DPT	20.7935	1.4372	2.0655
HMAX	27.8152	1.0579	1.1191
SLP	946.2696	1.0235	1.0475
TOC	2.4087	0.9521	0.9066

Table 5.4. Eigen values

NUMBER (Y_i)	Eigenvalue	Explained Var(%)	Accumulated
1	16041.3896	99.8942	99.8942
2	8.3492	0.052	99.9462
3	5.7647	0.0359	99.9821
4	1.1946	0.0074	99.9895
5	1.01	0.0063	99.9958
6	0.6476	0.004	99.9998
7	0.0248	0.0002	100

Average of the Eigenvalues = 2294.1

Table 5.5. Eigenvectors the first seven eigenvectors

No	e1	e2	e3	e4	e5	e6	e7
1	-0.0002	0.0365	-0.5913	-0.0491	0.0929	-0.1328	-0.7877
2	0.0095	0.9884	0.0636	0.0816	-0.0074	-0.11	0.0106

123

3	-0.0001	0.046	-0.7949	-0.0252	0.0191	-0.0081	0.6041
4	0.0009	0.11	-0.0071	-0.2335	0.585	0.768	-0.0356
5	-0.0002	-0.0859	0.0357	0.6702	0.673	-0.2936	0.0562
6	0	-0.0095	-0.0009	-0.0003	-0.0005	0.0005	-0.0002
7	0.0003	-0.0078	0.1145	-0.6976	0.4425	-0.5424	0.1009

Thus the PCs are expressed as:

$Y1 = -0.0002(DPT) + 0.0095(RH) + ... + 0.0003(SLP)$

$Y2 = 0.0365(DPT) + 0.9884(RH) + ... - 0.0078(SLP)$

$Y3 = -0.5913(DPT) + 0.0636(RH) + ... + 0.1145(SLP)$

$Y4 = -0.0491(DPT) + 0.0816(RH) + ... - 0.6976(SLP)$

$Y5 = 0.0929(DPT) - 0.0074(RH) + ... + 0.4425(SLP)$

$Y6 = -0.1328(DPT) - 0.1100(RH) + ... - 0.5424(SLP)$

$Y7 = -0.7877(DPT) + 0.0106(RH) + ... + 0.1009(SLP)$

Table 5.6. Scores of the Components

No	Y1	Y2	Y3	Y4	Y5	Y6	Y7
2	569.012	75.658	86.329	-637.457	439.722	-533.1	96.57
3	272.815	77.44	86.686	-635.726	439.326	-532.045	96.767
4	489.785	73.772	80.763	-635.462	439.301	-531.356	96.278
5	260.079	73.801	85.395	-636.453	440.188	-530.843	96.626
6	654.817	76.379	80.127	-636.965	439.915	-531.536	96.521
7	298.705	76.144	86.149	-634.802	440.655	-532.125	96.326
8	325.177	73.133	87.857	-636.482	440.454	-529.838	96.625
9	510.386	73.74	81.622	-637.681	440.825	-531.119	96.286
10	744.222	76.055	84.266	-634.433	440.269	-531.585	96.314
11	378.331	78.972	81.997	-637.502	441.936	-531.585	96.48
12	252.388	74.883	84.37	-637.616	440.888	-532.43	96.093

13	272.695	75.876	79.622	-635.959	441.545	-532.352	96.611
14	449.998	75.141	83.322	-636.686	439.212	-531.003	96.343
15	323.975	73.006	86.087	-636.464	439.425	-532.056	96.295
16	336.059	61.185	80.859	-636.571	439.976	-531.262	96.433
17	428.498	75.061	84.181	-636.629	438.662	-532.99	96.474
18	367.081	73.884	81.146	-636.285	440.398	-532.303	96.496
19	316.493	75.328	82.016	-636.992	439.695	-532.782	96.347
20	234.276	74.219	79.438	-635.932	438.386	-532.271	96.422
21	534.702	75.491	79.341	-635.856	440.131	-531.371	96.396
22	400.492	74.112	87.96	-635.542	442.112	-532.097	96.329
23	408.662	71.732	80.451	-637.277	440.534	-531.027	96.389
24	334.474	73.218	82.758	-637.876	440.049	-532.245	96.455
25	484.697	74.763	82.966	-636.26	442.26	-531.484	96.489
26	420.728	78.692	79.182	-637.452	440.651	-531.911	96.819
27	243.079	73.963	84.425	-637.333	439.872	-532.062	96.447
28	292.596	75.685	83.534	-637.837	439.908	-531.382	96.468
29	279.114	78.019	79.377	-635.083	439.732	-530.127	96.25
30	526.004	75.34	83.065	-637.14	441.351	-532.365	96.399
31	394.172	72.603	84.054	-638.79	440.094	-532.516	96.688
32	252.978	73.927	82.961	-636.366	440.628	-531.716	96.369
33	434.998	75.165	81.699	-635.816	441.187	-532.498	96.473
34	288.265	72.434	80.874	-634.892	441.422	-533.141	96.576
35	272.259	72.07	79.907	-637.605	442.527	-531.363	96.527
36	643.469	71.257	80.687	-635.766	441.257	-531.956	96.655
37	538.374	72.053	83.766	-636.478	440.286	-532.579	96.475
38	141.3	76.895	80.538	-634.979	440.537	-533.226	96.609
39	287.814	78.013	80.092	-637.811	438.907	-530.973	96.433
40	481.106	76.058	81.338	-639.3	438.83	-532.204	96.536
41	358.585	74.139	82.425	-638.093	442.515	-531.484	96.537
42	353.54	80.102	81.474	-635.416	440.727	-531.339	96.49

43	298.254	71.656	78.315	-635.567	439.646	-532.053	96.457
44	318.904	76.42	81.111	-638.096	441.653	-533.362	95.95
45	472.196	74.854	80.828	-636.407	440.54	-532.971	96.508
46	368.822	78.158	82.359	-637.597	441.075	-530.44	96.392

Table 5.7. Correlation variables x scores

Yi	DPT	RH	VP	TOC	HMAX	PYRF	SLP
Y_1	-0.0186	0.3876	-0.0096	0.1173	-0.0185	0	0.0381
Y_2	0.0733	0.9196	0.0694	0.3338	-0.2345	-.0002	-0.0223
Y_3	-0.9878	0.0492	-0.9961	-0.0178	0.0811	0	0.2685
Y_4	-0.0374	0.0287	-0.0144	-0.268	0.6924	0	-0.745
Y_5	0.0649	-.0024	0.01	0.6175	0.6394	0	0.4345
Y_6	-0.0744	-.0285	-0.0034	0.6491	-0.2234	0	-0.4264
Y_7	-0.0862	0.0005	0.0496	-0.0059	0.0084	0	0.0154

5.3.   S-CURVE

It is found that, single input vector input layer does not provide a good result during the training process and not trained as it should be. It is found that satisfactory input vectors provides trainable weights to construct relationship between independent variable (i.e., input vectors) and dependent variable (i.e, output).   Thus, single PC '$Y1$' cannot be applicable for the BPN model. Accordingly, four experiments are prepared, as presented in the Table, wherein model is trained by the one, two, three, and four PCs as input. Model is trained by 100 epochs for each experiment and obtained training error MAD (% of mean). A graph is plotted as depicted in the Fig. 5.1 for training and independent period.

For all experiment graphs characterizes **S-Curve** during training and independent period as depicted in the Fig.5.1. And, it is found that, two PCs (i.e., $Y1$, and $Y2$) input vectors have provided minimum MAD (% of mean) for training as well as independent period as shown in the Table 5.8 and Table 5.9 respectively.

Table. 5.8. MAD (% of mean) for different number of PCs as input during the training period, Epochs=100.

Number of PCs (n)	Experiment 1 (MAD % of mean)	Experiment 2 (MAD % of mean)	Experiment 3 (MAD % of mean)	Experiment 4 (MAD % of mean)
One ($Y1$)	4.07641657	4.076416578	4.076416578	4.076416578
Two ($Y1,Y2$)	4.05654193	4.051569022	4.050346762	4.047600512
Three ($Y1$, $Y2,Y5$)	4.07404977	4.06274567	4.075498748	4.07433457
Four ($Y1$, $Y2,Y5, Y7$)	4.07668288	4.051759913	4.07198419	4.064276279

Table. 5.9. MAD (% of mean) for different number of PCs as input during the independent period, Epochs=100

Number of PCs (n)	Experiment 1 (MAD % of mean)	Experiment 2 (MAD % of mean)	Experiment 3 (MAD % of mean)	Experiment 4 (MAD % of mean)
One ($Y1$)	3.768139243	3.728792908	3.715429116	3.782336688
Two ($Y1,Y2$)	3.692300222	3.682317072	3.680802414	3.688616996

Three ($Y1$, $Y2,Y5$)	3.721145956	3.68716207	3.705255296	3.717907059
Four ($Y1$, $Y2,Y5, Y7$)	3.698677665	3.657172138	3.699939828	3.675954787

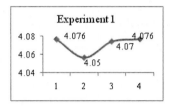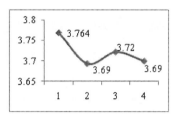

(a)

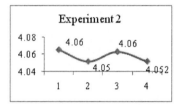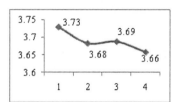

(b)

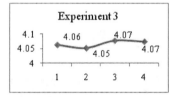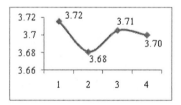

(c)

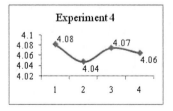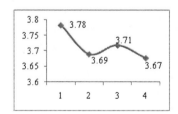

(d)

Fig. 5.1. Selection of number of PCs in input layer (a). Experiment 1 (b).

Experiment 2 (a). Experiment 3 (b). Experiment 4

As a result, it is found that, two PCs as input are acquired well-organized BPN model. Finally, two PCs as input vector, two neurons in hidden layer, one output neurons, sigmod function as neuron transfer function is used to develop PC-BPN model in parametric forecast and detail discussed in the following consequent section.

5.4.   PCs BPN MODEL

PC-BPN model is depicted in the following Fig. 5.2.   Wherein, two input PCs, Y_1 and $Y2$ are obtained from PCA is used to input instead of all seven parameters. Two neurons in hidden layer, one neuron in output layer, a total of 9 trainable weights including biases on hidden and one output layer in the three layers have been used to observe TMRF as model output. It has been observed that two neurons are best fitted in hidden layer therefore two neurons in hidden layer is preferred. The neuron output is obtained as $f(x_j)$. Where f is a transfer function typically the sigmoid function. The skeleton of the model is shown in Fig. 5.2 and Table 5.13.

Table 5.10. PCs-BPN Model.

Model input	Model type	LR	Neuron in Hidden Layer (p)	Input vector (n)	Trainable weights (w$_i$)	Target (y$_k$)
Principal Components 'Y1, and Y2 as Input	PCs-BPN	0.3	02	02	09 including biases	TMRF

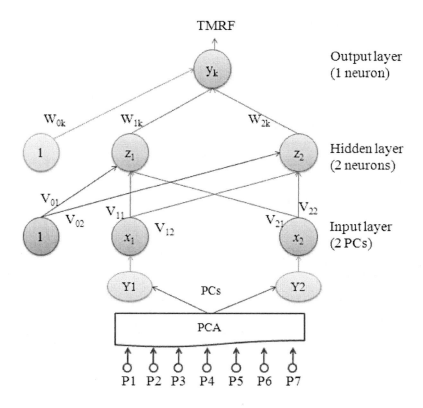

Fig. 5.2.PCs-BPN Model.

130

5.4.1. Training of the Model

Proposed algorithm 2.2 (Section 2.7) is used to train the PC-BPN. The initial random trainable weights including biases on hidden and output layer as shown in Table 5.11 is selected randomly by using "Random" class of Java. Model is trained to optimized weights by exercising 20, 00000 epochs. The optimized weights are shown in Table 5.12. The trained network presented excellent learning curve started with local minima MSE = 0.00105310182851675 to global minima MSE = 8.365814444121752E-04 that is exceptionally close to the optimum value of 0 as shown in the Table 5.13 and Fig. 5.3.

Table 5.11. Initial set of weights before training

Initial weight V_{ij}; $i = 1,2$; $j = 1,2$	
0.32798683643341064	0.46306312084198
0.1408836841583252	0.364432692527771
Initial weight $V0_{ij}$; $i = 1,2$	
0.9580116271972656	0.819179892539978
Initial weight W_i; $i = 1,2$	
0.07386475801467896	0.5617369413375854
Initial W0 : 0.5	

131

Table 5.12. Optimized set of weights before training, 20, 00000 epochs.

Optimized weights $V_{ij}; i = 1,2; j = 1,2$	
8.177974130768435	-11.832867119157655
-7.767316988370921	2.4476529021684255
Optimized weights $V0_{ij}; i = 1,2$	
3.311906842447251E-4	8.071556408826866E-4
Optimized weights $W_i; i = 1 to 3$	
4.779868917339107	5.609815216504555
Optimized W0 : 0.0023774778882080664	

Table 5.13. Training Results of Parametric ANN Model.

Time series	Epochs	Minimum MSE (Local Minima)	Final MSE (Global Minima)
TMRF	20,00000	0.0010531018285	8.36581444412E-04

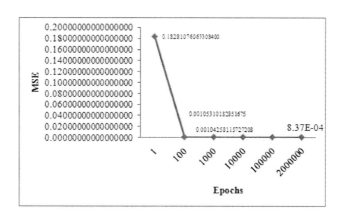

Fig. 5.3. Minimizing error (MSE) during training process.

132

5.4.2. Performance of the Model

PCs-BPN model in parametric forecast has been developed and verified independently. The performance of the model during the training period and independent/testing period is given in the Table 5.14 and 5.15 respectively. Wherein, the actual TMRF, predicted TMRF, and absolute deviation between them in the training as well as independent/testing period are presented. It is found that, the MAD (% of mean) is very less than the SD (% of mean) during the training period and also during the independent period as shown in Table 5.16. Correlation between actual and model predicted rainfall values is equal to 0.5 during the training period and 0.8 during the independent period. It is observed, model is moderately well for prediction of TMRF during the training period. It does not provide accuracy in the year of 1956, 1957, 1966, 1975, 1980, 1983, and 1991. The absolute deviation (im mm.) for these years also shows huge difference than SD i.e., 295.1 mm as depicted in the Fig 5.5. At the same time as, it shows good performance in the independent period. It shows poor performance in the year of 1994, 2001 and 2004. For other years it shows good performance. The absolute deviations (in mm.) are far below than the SD i.e., 353.7 mm in the independent period as depicted in the Fig.5.7 for all years apart from 1994 and 2001. In overall, it is found that model shows its significance in prediction however, not better evaluated over the model in parametric forecast (chapter 4). One of the causes is BPN model demands more input vectors as we have discussed in the chapter 2 instead of two only. To develop a relationship between dependent variable (output) and independent variables (inputs) the optimum numbers of trainable weights (adjustment weights) are always required. Two PCs dose not provides this requirement. One think have to keep in mind, opt for more parameters and apply PCA over them whenever would like to develop BPN model in parametric forecast. Generally, at least 8-10 PCs can produce enhanced performance.

Table. 5.14. Performance of the PCs-BPN model in training period (1953-1991).

Year	1	2	3	Actual TMRF (in mm.)	Predicted TMRF (in mm.)	Absolute Deviation (in mm)
1953	0.60690	0.61746	0.01055	1156.8	1246.5	89.7
1954	0.61983	0.62336	0.00353	1267.3	1298.8	31.5
1955	0.58123	0.60628	0.02505	957.6	1151.7	194.1
1956	0.69158	0.62862	0.06295	2049	1346.9	702.1
1957	0.58712	0.63397	0.04684	1001.1	1397.1	396.0
1958	0.61065	0.61391	0.00325	1188.1	1215.8	27.7
1959	0.60671	0.60864	0.00193	1155.2	1171.2	16.0
1960	0.63407	0.60870	0.02537	1398.1	1171.7	226.4
1961	0.67586	0.65316	0.02270	1848.2	1590.2	258.0
1962	0.58050	0.60154	0.02103	952.3	1113.1	160.8
1963	0.59860	0.63246	0.03385	1089.6	1382.7	293.1
1964	0.64675	0.62301	0.02374	1523.4	1295.6	227.8
1965	0.61510	0.60280	0.01229	1226	1123.3	102.7
1966	0.56075	0.60922	0.04846	815.4	1176.0	360.6
1967	0.59756	0.60805	0.01048	1081.4	1166.3	84.9
1968	0.57527	0.60112	0.02584	914.8	1109.7	194.9
1969	0.61127	0.60236	0.00891	1193.3	1119.7	73.6
1970	0.57402	0.61032	0.03629	906	1185.3	279.3
1971	0.66960	0.64186	0.02774	1773.5	1474.0	299.5

1972	0.61071	0.61180	0.00109	1188.6	1197.8	9.2
1973	0.60649	0.60123	0.00525	1153.4	1110.6	42.8
1974	0.57592	0.60197	0.02605	919.4	1116.5	197.1
1975	0.64784	0.60643	0.04141	1534.6	1152.9	381.7
1980	0.66087	0.60609	0.05478	1674	1150.1	523.9
1981	0.60033	0.60122	8.82E-4	1103.4	1110.5	7.1
1983	0.57837	0.63766	0.05929	936.9	1432.7	495.8
1984	0.60077	0.61539	0.01462	1106.9	1228.6	121.7
1985	0.62385	0.62068	0.00317	1303.2	1274.8	28.4
1986	0.61841	0.61022	0.00818	1254.8	1184.5	70.3
1987	0.61021	0.60192	0.00829	1184.4	1116.1	68.3
1988	0.61576	0.63252	0.01676	1231.7	1383.4	151.7
1989	0.61875	0.60119	0.01755	1257.8	1110.3	147.5
1990	0.64837	0.61845	0.02991	1540	1255.1	284.9
1991	0.65830	0.62450	0.03379	1645.6	1309.1	336.5

1-Actual Rainfall (Normalized), 2-Predicted TMRF (Normalized) and 3.
Absolute Deviation (Normalized)

Table.5.15. Performance of the PCs-BPN model in independent/testing period (1992-2004).

Year	1	2	3	Actual TMRF (in mm.)	Predicted TMRF (in mm.)	Absolute Deviation (in mm)
1992	0.61096	0.63420	0.02324	1190.7	1399.3	208.6
1993	0.62075	0.61422	0.00653	1275.5	1218.4	57.1
1994	0.69481	0.73030	0.03549	2092.8	2643.7	550.9
1995	0.60567	0.61864	0.01296	1146.7	1256.8	110.1

1998	0.59345	0.60522	0.01176	1049.3	1142.9	93.6
1999	0.61550	0.60430	0.01119	1229.5	1135.5	94.0
2000	0.61625	0.60382	0.01243	1236	1131.6	104.4
2001	0.67357	0.61591	0.05766	1820.5	1233.0	587.5
2002	0.59815	0.61021	0.01206	1086	1184.4	98.4
2003	0.61678	0.60482	0.01196	1240.6	1139.7	100.9
2004	0.56715	0.60258	0.03542	858.4	1121.5	263.1

1- Actual Rainfall (Normalized); 2- Predicted TMRF (Normalized); 3- Absolute Deviation (Normalized)

Table 5.16. Performance of PCs BPN model in parametric forecast during training and independent Period.

Data	Training period (1953-1991)					
	Mean	SD	MAD	SD (% of mean)	MAD (% of mean)	CC
Normalized	0.62	0.032	0.022	5.19	3.69	0.45
De-Normalized (in mm)	1251	295.1	202.5	23.6	16.2	0.45
Data	Independent/testing period (1992-2004)					
	Mean	SD	MAD	SD (% of mean)	MAD (% of mean)	CC
Normalized	0.62	0.036	0.021	5.77	3.38	0.8
De-Normalized (in mm)	1293.3	353.7	206.3	27.3	15.9	0.8

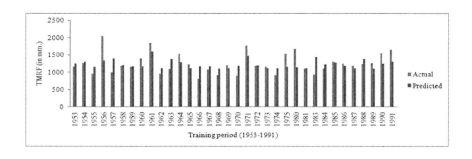

Fig. 5.4. Performance of PCs-BPN model during the training period
(1951-1991).

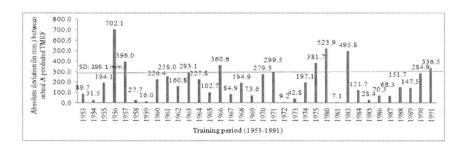

Fig. 5.5. Absolute deviation between actual TMRF and predicted TMRF
during training period (1951-1991).

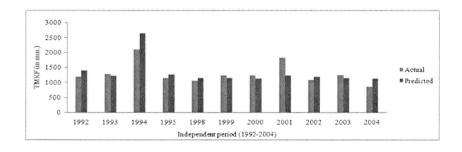

Fig. 5.6. Performance of PCs-BPN model during the independent/testing
period (1992-2004).

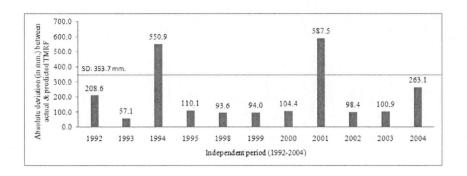

Fig. 5.7. Absolute deviation between actual TMRF and predicted TMFR during independent period/testing period.

5.5 CONCLUSIONS

PCs-BPN model have been developed for prediction of long-range TMRF over the smaller region and assessed in the training as well as independent period. The entire development process of the model is discussed in this chapter. It is observed that the model can perform well for pattern recognition as well as prediction however the performance is entirely depending on selection of number of input PCs. As we know that sufficient number of input parameters are required to obtain better result instead of two only. It is concluded that, this technique is very useful if large number of input parameters is used. As we also know that large input vectors will introduce unknown trainable weights in the network as well. Thus through PCA, we can reduce correlated parameters and can obtained PCs to be used. This is only one solution to reduce unknown trainable weights. Thus this model is significant and relevant for the study. Here, developed Java based simulator for this model is given in the following Figure 5.8.

138

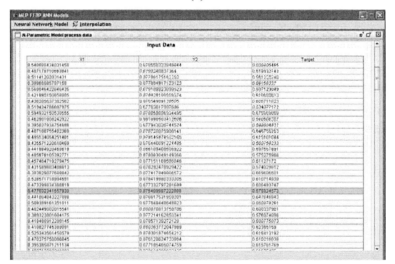

(a)

(b)

139

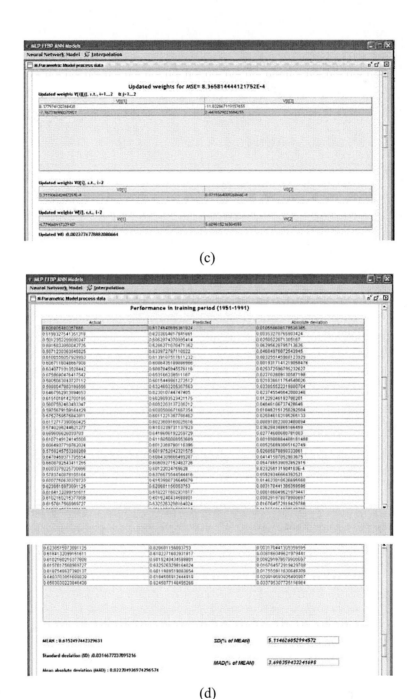

(c)

(d)

Fig. 5.8. Java based PCs BPN model output screens.

Chapter 6

HYBRID BPN MODEL

A hybrid BPN model is developed and evaluated. Wherein, output of all models (i.e. deterministic, parametric, and PCs) are provided as input to new BPN model and trained. The performance of model is evaluated and also compared with all individual models. It is found that the performance of the model is satisfactory and may be utilize BPN in this manner as well is discussed in this chapter.

6.1. INTRODUCTION

A special class of application of BPN named as Hybrid BPN model is discussed in this chapter. It is observed that BPN in model parametric forecast is better evaluated over the PBN model deterministic forecast as well as PC-BPN model in parametric forecast. A hybrid BPN model is introduced, wherein output of all three models has been used as input of a new BPN model to observed TMRF as an output. Actual benefit of this architecture is that, we may able to more trainable weights in the network to develop relationship between dependent and independent variables. In the previous model we have seen that break-down situations. It means that we cannot introduce more trainable weights. In this model 40 weights from deterministic forecast, 31 weights from parametric forecast, 9 trainable weights from PC based parametric forest is used to develop relationship. Two neurons in hidden layer, three input vectors in input layer are used in new BPN. Thus total of 11 trainable weights including biases plus 40, 31, 9 trained weights have been used in this hybrid model. The performance of the hybrid model is presented in this chapter as well.

6.2. HYBRID BPN MODEL

The architecture of the model is depicted as following Fig. 6.1 wherein, predicted TMRF R1, R2, and R3 from deterministic BPN model, BPN model in parametric forecast, and PC BPN model respectively are inputted to another BPN model to observed TMRF. Two neurons in hidden layer, one neuron in output layer are used. Thus total trainable weights included biases 40+31+9+11 = 91 are used in this model.

6.2.1. Training of the Model

Proposed algorithm 2.2 (Section 2.7) is used to train the model. The initial random trainable weights including biases on hidden and output layer as shown in Table 6.1 is selected randomly by using "Random" class of Java. Model is trained to optimized weights by exercising 20, 00000 epochs. The optimized weights are shown in Table 6.2. The trained network presented excellent learning curve started with local minima MSE = 8.62542961803220E-04 obtained at 1000 epochs to global minima MSE = 3.812664778766369E-05 obtained at 20,00000 epochs that is exceptionally very closed to the optimum value of zero as shown in the Table 6.3., and depicted in the Fig. 6.2.

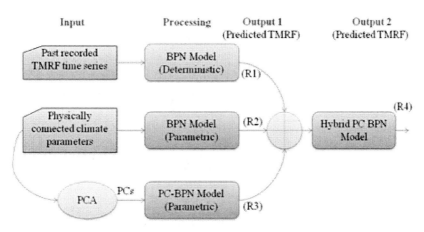

Fig. 6.1. Hybrid BPN Model

143

Table 6.1. Initial set of weights before training.

Initial weight V_{ij}; $i = 1,2,3$; $j = 1,2$	
0.7764207720756531	0.42187821865081787
0.7362191081047058	0.8560808897018433
0.6567222476005554	0.7526206970214844
Initial weight $V0_{ij}$; $i = 1,2$	
0.6463848352432251	0.5990971326828003
Initial weight W_i; $i = 1,2$	
0.62311851978302	0.5429116487503052
Initial W0 : 0.5	

Table 6.2. Optimized set of weights before training, 20, 00000 epochs.

Optimized weight V_{ij}; $i = 1,2,3$; $j = 1,2$	
3.268283378360706	-1.9266929295914503
0.3546624007478421	-0.00884791125128299
-1.9907203010853274	-1.0263075979195142
Optimized weight $V0_{ij}$; $i = 1,2$	
-2.2280082591129475E-4	5.381164356498577E-4
Optimized weight W_i; $i = 1,2$	
2.646505921511638	-10.489241704787911
Optimized W0 : -4.634714556271986E-04	

144

Table 6.3. Training Results.

Time series	Epochs	Minimum MSE (Local Minima)	Final MSE (Global Minima)
TMRF	20,00000	8.6254296180E-04	3.812664778766E-05

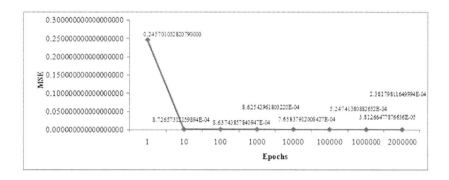

Fig. 6.2. Learning Curve

6.2.2. Performance of the Model

Hybrid BPN model has been developed and verified independently. The performance of the model during the training period and independent/testing period is given in the Table 6.4 and Table 6.5 respectively. Wherein, the actual TMRF, predicted TMRF, and absolute deviation between them in the training as well as independent/testing period are presented. It is found that, the MAD (% of mean) is very less than the SD (% of mean) during the training period however not in the independent period as shown in the Table 6.5. The model trained superbly, the high correlation (i.e., 0.97) is obtained between actual and predicted TMRF. The absolute deviations between actual and predicted

145

are shown in Fig. 6.3 and 6.4 are shown unbelievably excellent performance in the training period. However, moderate performance has been found during the independent/testing period. Moderate correlation i.e., 0.5 has been found between the actual and predicted TMRF by eliminating unpredictable independent year 1994 and 2004 as shown in the Table 6.5 and 6.6. And also it is observed that, the SD (% of mean) i.e., 18.02 is greater to MAD (% of mean) i.e., 14.6. These facts speak s the model is completely undesirable. Model is unable to explain the TMRF for the year of 1994, 2001, and 2004 (Fig. 6.5). Besides, model successfully explains the variability for the others years (Fig. 6.6).

Table. 6.4. Performance of the Hybrid BPN in training period (1962-1991)

Year	1	2	3	Actual TMRF (in mm.)	Predicted TMRF (in mm.)	Absolute Deviation (in mm)
1962	0.58050	0.57559	0.00491	952.3	917.0	35.3
1963	0.59860	0.60543	0.00682	1089.6	1144.7	55.1
1964	0.64675	0.64527	0.00148	1523.4	1508.3	15.1
1965	0.61510	0.61985	0.00475	1226	1267.5	41.5
1966	0.56075	0.55775	0.00300	815.4	795.6	19.8
1967	0.59756	0.60467	0.00710	1081.4	1138.4	57.0
1968	0.57527	0.58109	0.00581	914.8	956.6	41.8
1969	0.61127	0.61282	0.00155	1193.3	1206.5	13.2
1970	0.57402	0.57934	0.00531	906	943.9	37.9
1971	0.66960	0.66834	0.00125	1773.5	1758.8	14.7
1972	0.61071	0.61530	0.00458	1188.6	1227.7	39.1
1973	0.60649	0.61273	0.00624	1153.4	1205.8	52.4

1974	0.57592	0.57542	5.00E-4	919.4	915.9	3.5
1975	0.64784	0.64438	0.00346	1534.6	1499.3	35.3
1980	0.66087	0.65904	0.00183	1674	1653.7	20.3
1981	0.60033	0.60473	4.40E-3	1103.4	1139.0	35.6
1983	0.57837	0.57944	0.00107	936.9	944.6	7.7
1984	0.60077	0.60288	0.00210	1106.9	1123.9	17.0
1985	0.62385	0.61402	0.00982	1303.2	1216.7	86.5
1986	0.61841	0.62148	0.00307	1254.8	1282.0	27.2
1987	0.61021	0.59283	0.01737	1184.4	1044.5	139.9
1988	0.61576	0.61165	0.00410	1231.7	1196.6	35.1
1989	0.61875	0.61760	0.00115	1257.8	1247.7	10.1
1990	0.64837	0.63746	0.01090	1540	1430.7	109.3
1991	0.65830	0.66524	0.00693	1645.6	1723.1	77.5

1- Actual Rainfall (Normalized); 2- Predicted TMRF (Normalized)

3- Absolute Deviation (Normalized)

Table.6.5. Performance of the hybrid BPN model in independent/testing period (1992-2004).

Year	1	2	3	Actual TMRF (in mm.)	Predicted TMRF (in mm.)	Absolute Deviation (in mm)
1992	0.61096	0.61924	0.00827	1190.7	1262.1	71.4
1993	0.62075	0.64709	0.02633	1275.5	1526.8	251.3
1994	**0.69481**	**0.60488**	**0.08992**	**2092.8**	**1140.2**	**952.6**
1995	0.60567	0.60118	0.00449	1146.7	1110.2	36.5
1998	0.59345	0.59837	0.00492	1049.3	1087.8	38.5
1999	0.61550	0.60648	0.00902	1229.5	1153.3	76.2

2000	0.61625	0.59457	0.02168	1236	1058.0	178.0
2001	0.67357	0.61591	0.05766	1820.5	1233.0	587.5
2002	0.59815	0.61021	0.01206	1086	1184.4	98.4
2003	0.61678	0.60482	0.01196	1240.6	1139.7	100.9
2004	**0.56715**	**0.64974**	**0.08258**	**858.4**	**1554.2**	**695.8**

1- Actual Rainfall (Normalized); 2- Predicted TMRF (Normalized)

3- Absolute Deviation (Normalized)

Table 6.6. Performance of hybrid BPN during training and independent Period.

Data	Training period (1951-1991)					
	Mean	SD	MAD	SD (% of mean)	MAD (% of mean)	CC
Normalized	0.61	0.03	0.005	4.9	0.78	0.97
De-Normalized (in mm)	1220.4	263.3	41.1	21.6	3.37	0.97
Data	Independent/testing period (1991-2004)					
	Mean	SD	MAD	SD (% of mean)	MAD (% of mean)	CC
Normalized	0.62	0.04	0.033	5.8	5.3	0.5
De-Normalized (in mm)	1293.2	353.7	300.2	27.3	23.2	0.5

148

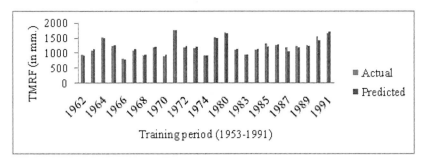

Fig. 6.3. Performance of hybrid BPN during the training period (1962-1991).

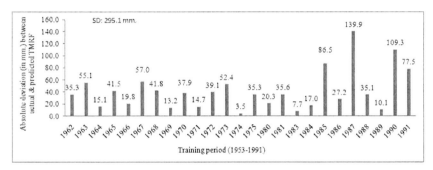

Fig. 6.4. Absolute deviation between actual TMRF and predicted TMRF during training period (1951-1991).

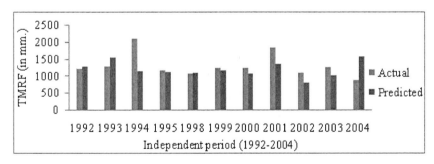

Fig. 6.5. Performance of hybrid BPN during the independent/testing period (1992-2004).

149

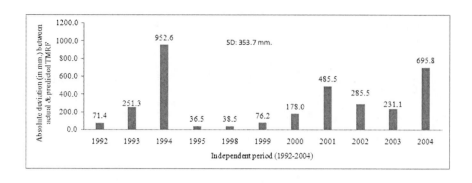

Fig. 6.6. Absolute deviation between actual TMRF and predicted TMFR during independent period/testing period.

6.3.   CONCLUSIONS

Deterministic as well as parametric BPN models has been hybridized and obtained Hybrid BPN model for prediction of long-range TMRF over the district and assessed in the training as well as independent period. The entire development process has been discussed in this chapter. It has been observed that the hybridizing in such a manner is not healthy for pattern recognition as well as prediction for this application. In general, it is concluded that, hybridization of parametric and deterministic model in this application may be useful by utilize different set of input data in identification of internal dynamics of TMRF over the district however; not a better solution can declare especially in this case. It is concluded that the researcher must go for either deterministic or parametric way. If they able to identify large number input parameters then they must go for PCA to obtain PCs and than may use parametric model for better solution. Here, developed Java based simulator for hybrid model is given in the following Figure 6.7. Output screens are shown here for 2000000 epochs.

150

(a)

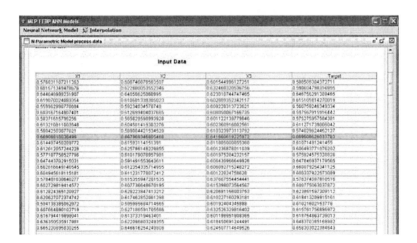

(b)

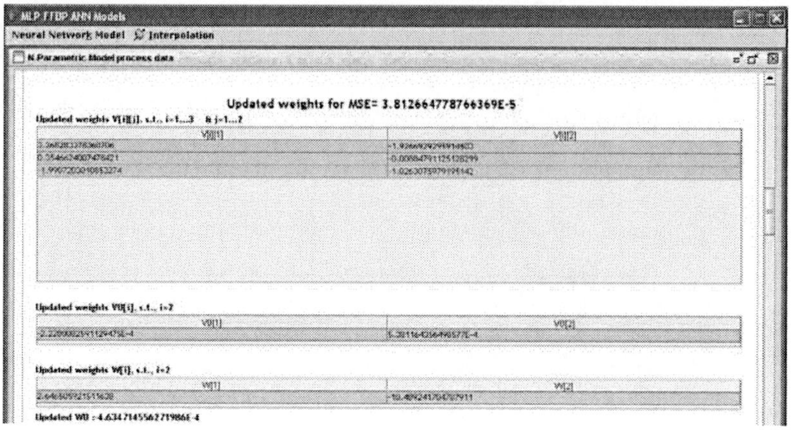
MLP FFBP ANN Models
Neural Network Model
Interpolation
N-Parametric Model process data
Updated weights for MSE= 3.812664778766369E-5

(c)

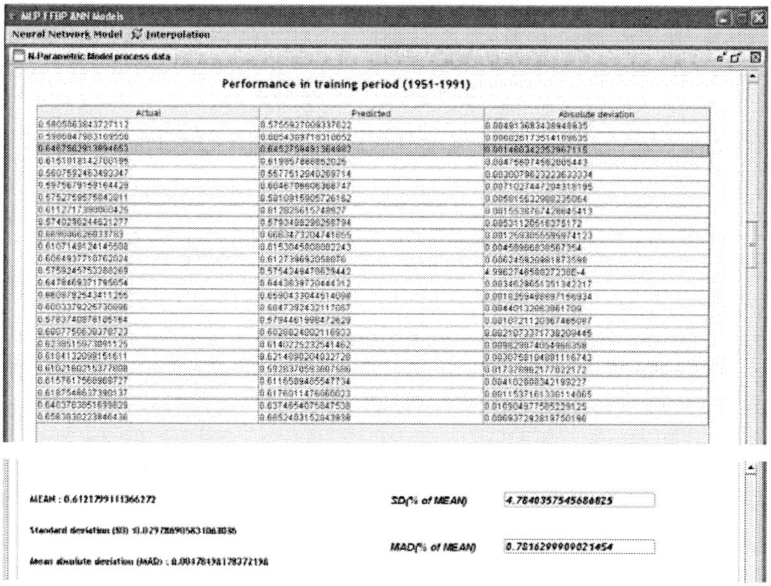
MLP FFBP ANN Models
Neural Network Model
Interpolation
N-Parametric Model process data
Performance in training period (1951-1991)
Actual
Predicted
Absolute deviation
MEAN : 0.6121799111366272
SD(% of MEAN)
4.784035754568 6825
MAD(% of MEAN)
0.7816299909021454

(d)

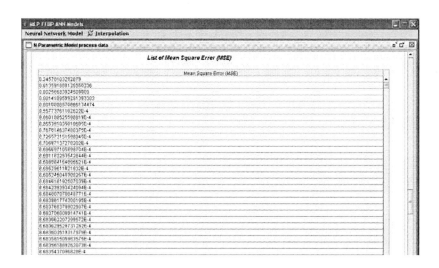

(e)

(f)

Fig. 6.7. Java based Hybid BPN model output screens.

Chapter 7

SUMMARY AND CONCLUSIONS

This chapter presents the summary of developed models for long-range monsoon rainfall over the smaller region. Here, emphasis has been given to conclusions of the work.

7.1.   SUMMARY

Long-range monsoon rainfall prediction over a smaller region like districts or subdivisions is very complicated. Since 1999, India Meteorology Department, Pune, has been issuing long-range forecasts for three broad homogeneous regions of India, viz., Northwest India, Northeast India, and the Peninsular, using statistical power regression models. However, it has been found that the performances of such models are not satisfactory and they fail to study the highly non-linear relationships between climate and its predictors over the districts or subdivisions. Broad literature from 1923 to 2011 has been reviewed (Chapter 1). It has been found that the ANN model may conquer this restriction. Therefore, we have projected a subject of this MRP work is the development of the ANN model for long-range weather forecasting over a smaller scale geographical region like district in the context of Chhattisgarh region. For that reason, four BPN models have been developed and evaluated. Models are developed by using Java language (Chapter 2). These models are-

1. BPN model in deterministic forecast (Chapter 3).
2. BPN model in parametric forecast (Chapter 4).
3. PCs-BPN model (Chapter 5).
4. Hybrid BPN model (Chapter 6).

7.1.1.   BPN model in deterministic forecast

Developed BPN model in deterministic forecast is depicted in Fig. 7.1. The parameters of the final model are as follows. It is concluded

156

that model is useful for prediction of long-range monsoon rainfall over the very smaller region like district. It is capable to explain the internal dynamics of long-range monsoon rainfall from only it's past recorded data time series. However, it is found that this not capable to explain the surplus rainfall in the year of 1994 and shortfall rainfall in the year of 2004.

1.	Name of the model	:	BPN model in deterministic forecast.
2.	Number of layer	:	03
3.	Input vector (n)	:	11 inputs (11 years TMRF data time series)
4.	Input data (independent parameter)	:	11 years TMRF data time series.
5.	Output neurons	:	One neuron.
6.	Target (Dependent parameter)	:	12th year TMRF.
7.	Hidden layer	:	01
8.	Neurons in hidden layer (p)	:	03
9.	Transfer function (f)	:	Sigmoid.
10.	Learning rate (α)	:	0.3
11.	Weight optimization rule	:	Delta learning rule (Rumelhart et $al.$, 1986, Back-propagation algorithm).
12.	Number of weights (w_i)	:	40 (Including biases).
13.	Training algorithm	:	Algorithm 2.1, Section 2.7. Chapter

2.

14.	Epochs	:	20,00000
15.	Local Minima-MSE (M_L)	:	8.6719891654087650E-04
16.	Global minima-MSE (M_G)	:	3.8712559948953136E-05
17.	Trained weights	:	Table 7.1.
18.	Performance	:	Excellent in training period, and good in the independent period Over all Good as shown in Table 7.2 and Figure 7.2.

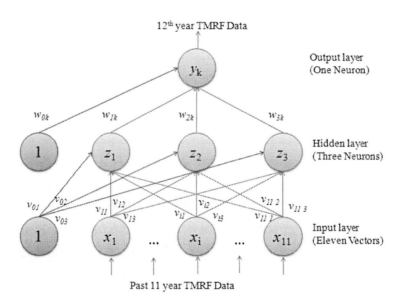

Fig. 7.1. BPN in deterministic forecast.

Table 7.1. Trained weights.

Optimized weight $V_{ij}; i = 1,2,3,...,11; j = 1,2,3$		
6.270598280340316	-2.0018271973656407	1.6423065173514284
-3.9865007440307214	11.079553010431491	7.553871579443576
0.199126014942202	2.6061153786907916	3.7054459154149
1.8663610907790922	-10.234867498568144	-5.206031833333133
-1.3285773882910707	-1.6445481520677703	-1.4167666469550697
-2.2955962777155356	0.10258439352523269	-0.6287308576104448
-5.060791432018832	-0.5036163178883608	-5.034922160467127
6.030420540993752	2.250903868798964	4.040110957750795
-3.6506887651307047	-3.3864106511729157	-3.86532590441335
-3.3102116272578717	2.245673840960485	-1.1181211626486898
8.612823408332893	-3.711303665382227	3.0618059057010174
Optimized weight $V0_i; i = 1,2,3$		
-2.71273854670431E-4	-5.818654669719801E-4	2.9776089214191855E-4
Optimized weight $W_i; i = 1,2,3$		
6.731181177946909	7.43777627223151	-7.711594995180171
Optimized W0 : -4.6218733807633356E-04		

Table 7.2. Performance of BPN in deterministic forecast during training (1951-2004) and independent Period (1992-2004).

Data	Training period (1951-1991)					
	Mean	SD	MAD	SD (% of mean)	MAD (% of mean)	CC
Normalized	0.61	0.034	0.005	4.78	0.76	0.98
De-Normalized (in mm)	1220.4	263.3	39.9	21.6	3.3	0.98
Data	Independent/testing period (1991-2004)					
	Mean	SD	MAD	SD (% of mean)	MAD (% of mean)	CC
Normalized	0.62	0.034	0.04	5.50	6.41	0.5
De-Normalized (in mm)	1293.3	353.8	312.0	27.4	24.1	0.5

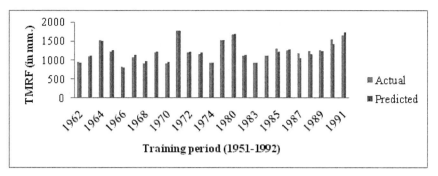

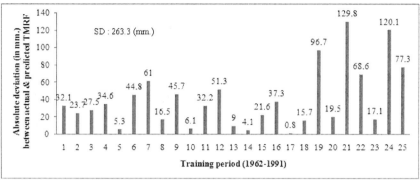

(a) Training period.

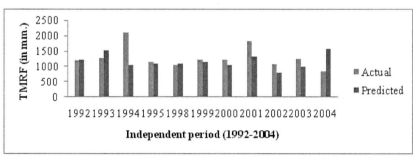

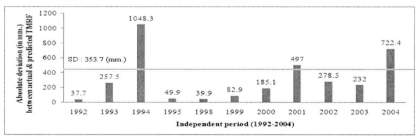

(b) Independent period.

161

Fig. 7.2. Performance of BPN model in deterministic forecast during training and independent period.

7.1.2.   BPN model in parametric forecast

Developed BPN model in parametric forecast is depicted in Fig. 7.3. The parameters of the final model are as follows. It is found that, identification of climate variable also knows as predictors (i.e., independent parameters) this are physically connected with the monsoon rainfall over the proposed area is a very difficult task. However, some predictors are successfully identified of this region as shown in Table 7.3. It is concluded that model is useful for prediction of long-range monsoon rainfall over the very smaller region. It is capable to explain the internal dynamics of long-range monsoon rainfall from only its predictor's data time series. it is found that this model some extent explains the surplus rainfall in the year of 1994 and successfully explained shortfall rainfall in the year of 2004. It is better evaluated over the deterministic model.

1.	Name of the model	:	BPN model in parametric forecast.
2.	Number of layer	:	03
3.	Input vector (*n*)	:	8
4.	Input data (independent parameter)	:	8 predictors data time series.
5.	Output neurons	:	One neuron.
6.	Target (Dependent parameter)	:	TMRF (same year)
7.	Hidden layer	:	01
8.	Neurons in hidden	:	03

layer   (*p*)

9.	Transfer function (*f*)	:	Sigmoid.
10.	Learning rate (α)	:	0.3
11.	Weight optimization rule	:	Delta learning rule (Rumelhart *et al.*, 1986, Back-propagation algorithm).
12.	Number of weights (w_i)	:	27 (Including biases).
13.	Training algorithm	:	Algorithm 2.2, Section 2.7. Chapter 2.
14.	Epochs	:	20,00000
15.	Local Minima-MSE (M_L)	:	0.0010583407253476
16.	Global minima-MSE (M_G)	:	6.555600546440187E-04
17.	Trained weights	:	Table 7.4.
18.	Performance	:	Excellent in training period, and very good in the independent period Over all outstanding Table 7.5 and Figure 7.4.

163

Table 7.3. Identified predictors for BPN model in parametric forecast.

No.	Parameter	Abb.	Month	Correlation coefficient (Period: 1981-2004)
P1	Mean Wet Bulb Temperature (in Deg. C)	WBT	Jun. (Prev. year 12 HR)	0.41
P2	Mean Dew Point Temperature (in Deg. C)	DPT	Jun. (Prev. year 03 HR)	0.45
P3	Mean Relative Humidity (in %)	RH	Aug. (Prev. year 03 HR)	-0.45
P4	Mean Vapour Pressure (in hPa)	VP	Jun. (Prev. year 03 HR)	0.55
P5	Mean Total Cloud Amount (in Oktas)	TOC	Oct. (Prev. year 03 HR)	-0.44
P6	Mean Station Level Pressure (in hPa)	SLP	Feb. (Current year 03 HR)	-0.41
P7	Highest Maximum Temperature (in Deg. C)	HMAX	Jan. (Current year)	0.56
P8	Previous year rainfall (in mm.)	PYRF	Jun. (Prev. year)	-0.42

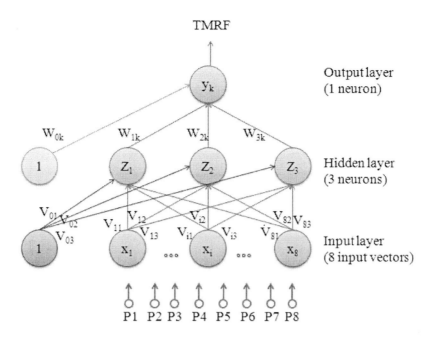

Fig. 7.3. BPN Model in parametric forecast.

Table 7.4. Trained weights.

Optimized weights V_{ij}; $i = 1,2,3,...8$; $j = 1,2,3$		
1.1504005495726712	-3.231188742943418	-4.49444937122248
8.145844471279098	-9.468212927557897	1.626150240201612
0.7304107553183267	-3.5220397850993557	-9.165839546716056
1.2818430196861041	-1.653726284147089	18.101521791435896
6.2718139989280415	10.830371711169267	-1.3520536437384232
-12.26532261361735	13.31525241952504	6.86774036208046
-2.7422444068742533	4.899050852156461	-8.019181173086945
0.17076326117063892	-2.629295308689212	-8.920492820789686
Optimized weights $V0_{ij}$; $i = 1,2,3$		
-3.79105590770616E-4	2.75069112419501E-4	5.949836614418309E-4

165

Optimized weights $W_i; i = 1,2,3$		
-2.6378943872281324	2.2084526467128085	7.2276776719864335
W0 : 8.011408367628631E-04		

Table 7.5. Performance of BPN in parametric forecast during training and independent Period

Data	Training period (1952-1991)					
	Mean	SD	MAD	SD (% of mean)	MAD (% of mean)	CC
Normalized	0.62	0.03	0.02	5.51	3.21	0.6
De-Normalized (in mm)	1257.8	293.6	174.7	23.3	13.8	0.6
Data	Independent/testing period (1992-2004)					
	Mean	SD	MAD	SD (% of mean)	MAD (% of mean)	CC
Normalized	0.62	0.03	0.02	5.50	2.96	0.8
De-Normalized (in mm)	1293.2	353.7	188.2	27.4	14.5	0.8

166

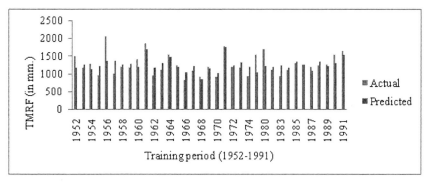

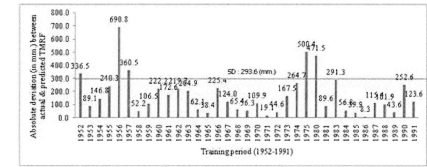

(a). Training period.

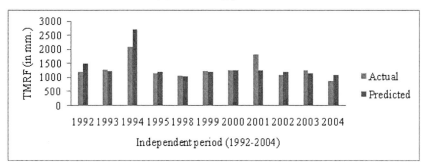

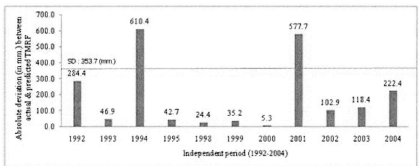

(b) Independent period.

Fig. 7.4. Performance of the BPN model in parametric forecast during training and independent period.

7.1.3. PCs-BPN model

Developed PCs-BPN model is depicted in Fig. 7.5. The parameters of the final model are as follows. PCA have been accomplished over the 7 predictor's parameter and obtain two PCs. These two PCs are used as input to the model and observed TMRF. It is found that, this model explains the surplus rainfall in the year of 1994 and successfully explained shortfall rainfall in the year of 2004 as BPN model in parametric forecast. Similar performance has shown by this model as well however, far better than the deterministic model. It is also noted that this model may perform better as compared to the parametric model if we are capable to produce more PCs. For that, we have to search out more predictors (i.e., independent variables those are physically connected to monsoon rainfall. Though, this is a very hectic task.

1. Name of the model : PCs-BPN model.
2. Number of layer : 03
3. Input vector (n) : 2
4. Input data (independent : 2 PCs data time series.
   parameter)
5. Output neurons : One neuron.
6. Target (Dependent : TMRF (same year).
   parameter)

168

7.	Hidden layer	:	01
8.	Neurons in hidden layer (p)	:	02
9.	Transfer function (f)	:	Sigmoid.
10.	Learning rate (α)	:	0.3
11.	Weight optimization rule	:	Delta learning rule (Rumelhart *et al.*, 1986, Back-propagation algorithm).
12.	Number of weights (w_i)	:	09 (Including biases).
13.	Training algorithm	:	Algorithm 2.2, Section 2.7. Chapter 2.
14.	Epochs	:	20,00000
15.	Local Minima-MSE (M_L)	:	0.00105310182851675
16.	Global minima-MSE (M_G)	:	8.365814444121752E-04
17.	Trained weights	:	Table 7.6.
18.	Performance	:	Excellent in training period, and very good in the independent period Over all outstanding Table 7.7 and Figure 7.6.

Table 7.6. Trained weights.

Initial weight V_{ij}; $i = 1,2$; $j = 1,2$	
8.177974130768435	-11.832867119157655

169

-7.767316988370921	2.4476529021684255
Initial weight $V0_{ij}; i = 1,2$	
3.3119068424472511E-4	8.071556408826866E-4
Initial weight $W_i; i = 1,2$	
4.779868917339107	5.609815216504555
Initial W0 : 0.0023774778882080664	

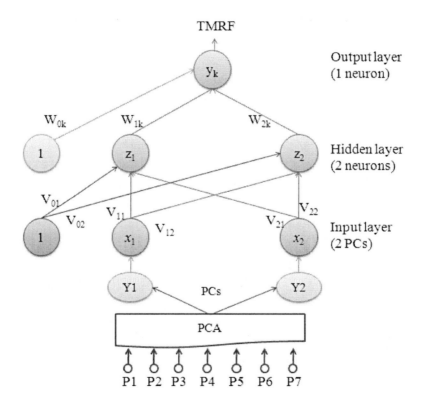

Fig. 7.5. PCs-BPN Model.

170

Table 7.7. Performance of PCs -BPN model in parametric forecast during training and independent Period.

Data	Training period (1953-1991)					
	Mean	SD	MAD	SD (% of mean)	MAD (% of mean)	CC
Normalized	0.62	0.032	0.022	5.19	3.69	0.45
De-Normalized (in mm)	1251	295.1	202.5	23.6	16.2	0.45
Data	Independent/testing period (1992-2004)					
	Mean	SD	MAD	SD (% of mean)	MAD (% of mean)	CC
Normalized	0.62	0.036	0.021	5.77	3.38	0.8
De-Normalized (in mm)	1293.3	353.7	206.3	27.3	15.9	0.8

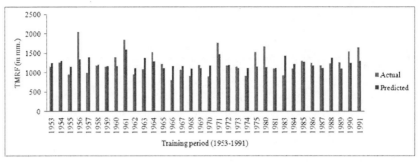

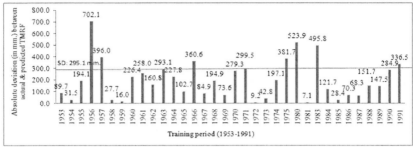

(a) Training period.

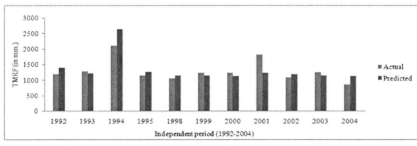

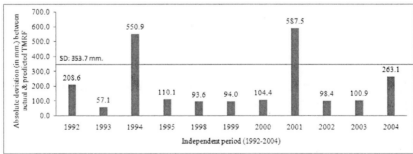

(b) Independent period.

Fig. 7.6. Performance of the PCs-BPN model during training and independent period.

7.1.4. Hybrid BPN model

Developed Hybrid BPN model is depicted in Fig. 7.7. The parameters of the final model are as follows. PCA have been accomplished over the 7 predictor's parameter and obtain two PCs. These two PCs are used as input to the model and observed TMRF. Model successfully trained however provided poor result as compared to parametric as well principal component based model. It is also observed that the implementation of this model is very complex however not provided result as had desired.

1.	Name of the model	:	Hybrid BPN model.
2.	Number of layer	:	03
3.	Input vector (n)	:	2
4.	Input data (independent parameter)	:	Output of deterministic, parametric, and Principal component models.
5.	Output neurons	:	One neuron.
6.	Target (Dependent parameter)	:	TMRF (same year).
7.	Hidden layer	:	01
8.	Neurons in hidden layer (p)	:	02
9.	Transfer function (f)	:	Sigmoid.
10.	Learning rate (α)	:	0.3

11.	Weight optimization rule	:	Delta learning rule (Rumelhart *et al.*, 1986, Back-propagation algorithm).
12.	Number of weights (w_i)	:	11 (Including biases) plus 40, 31, 9 trained weights from deterministic, parametric, and PCs based model.
13.	Training algorithm	:	Algorithm 2.2, Section 2.7. Chapter 2.
14.	Epochs	:	20,00000
15.	Local Minima-MSE (M_L)	:	8.62542961803220E-04
16.	Global minima-MSE (M_G)	:	3.812664778766369E-05
17.	Trained weights	:	Table 7.6.
18.	Performance	:	Excellent in training period, and poor in the independent period Over all poor performance as shown in the Table 7.7 and Figure 7.8.

Table 7.6. Trained weights.

Optimized weight $V_{ij}; i = 1,2,3; j = 1,2$	
3.268283378360706	-1.9266929295914503
0.3546624007478421	-0.00884791125128299
-1.9907203010853274	-1.0263075979195142

Optimized weight $V0_{ij}\,; i = 1,2$	
-2.2280082591129475E-4	5.381164356498577E-4
Optimized weight $W_i\,; i = 1,2$	
2.646505921511638	-10.489241704787911
Optimized W0 : -4.634714556271986E-04	

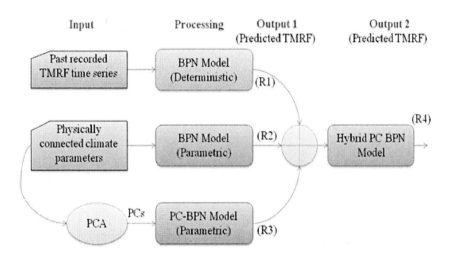

Fig. 7.7. Hybrid BPN Model.

Table 7.7. Performance of hybrid BPN model in parametric forecast
during training and independent Period.

Data	Training period (1962-1991)					
	Mean	SD	MAD	SD (% of mean)	MAD (% of mean)	CC

175

	Mean	SD	MAD	SD (% of mean)	MAD (% of mean)	CC
Normalized	0.61	0.03	0.005	4.9	0.78	0.97
De-Normalized (in mm)	1220.4	263.3	41.1	21.6	3.37	0.97
Data	Independent/testing period (1992-2004)					
	Mean	SD	MAD	SD (% of mean)	MAD (% of mean)	CC
Normalized	0.62	0.04	0.033	5.8	5.3	0.5
De-Normalized (in mm)	1293.2	353.7	300.2	27.3	23.2	0.5

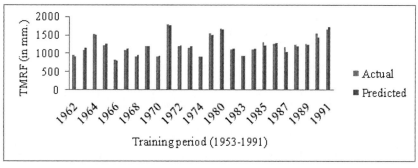

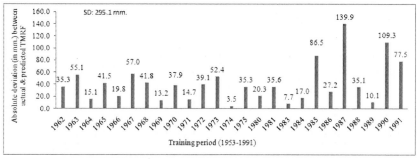

(a) Training period.

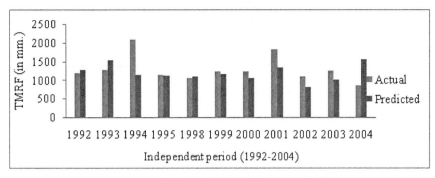

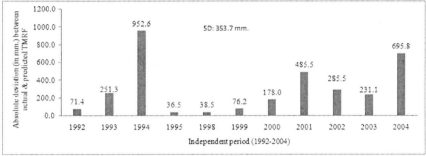

(b) Independent period

Fig. 7.8. Performance of hybrid BPN model during training and independent period.

7.2. CONCLUSIONS

The following conclusions are being drawn along with the results in a summarized form:

1. Long-range monsoon rainfall data time series over a smaller geographical region behave as chaotic motion. In short it represents

177

highly non-linear feature and sensitive to initial conditions. Hence they are very difficult to predict.

2.   There are two ways of prediction could be made. One based on the dynamical atmospheric prediction and the other through empirical/statistical prediction.

3.   Through dynamical atmospheric general circulation models with specified boundary conditions and varying initial conditions, as of today, globally dynamical models do not have the required skill to accurately simulate the salient features of the mean monsoon and weather and its variability. They are further hampered by the lack of data in the oceanic regions important to the climate.

4.   Two main drawbacks of statistical models have been observed:

    i.   Statistical models are not useful to study the highly nonlinear relationships between rainfall and its predictors.

    ii.   There is no ultimate end in finding the best predictors. It will never be possible to get different sets of regional and global predictors to explain the variability of the two neighboring regions having distinguished rainfall features.

5.   The ANN technique is able to get rid of the above two drawbacks

6.   There are many types of neural network viz., CPNN, RBFN, CCNN, PNN, GRNN etc. However, BPN model have been proposed and implemented by using Java.

7.   BPN is chosen for three reasons.

    a.   First, a lot of studies have been made on BPN and their abilities, especially in the case of chaos prediction.

b. Second, the availability of meteorological data corresponds to the needs of BPN with input and output data. And third, its use and its construction and development are easy.

8. For training of the models, algorithm 2.1 and 2.2 are proposed and implemented. In both the algorithm, BPN concept is used to feed forward and back propagation process during training period. Being a gradient descent algorithm, it minimizes MSE of the models. Although algorithm converges very slowly, however models are nicely trained for all cases.

9. There is no automatic way to select neural network parameters, and also development of a neural network and its training for particular weather data time series is a challenging task, however, it is observed that the relation between MAD and number of neurons in hidden layer (p) is $MAD \propto p$ and for number of input vector (n) is $MAD \propto \dfrac{1}{n}$. Finally it is observed that number of neurons in hidden layer $p = 2$ or 3 and number of input vectors $n = 11$ or more offered optimum results for all weather data time series. This configuration has produced minimum network MSE very close to the optimum value of 0.

10. A simple suggestion is to increase the learning rate in order to improve performance of the model and to decrease the learning rate in order to worsen the performance of model. Learning rate $\alpha = 0.3$ is selected for all of the cases.

11. For nearly all problems, one hidden layer is sufficient. There is no theoretical reason for using more than two hidden layers. It is

observed that number of hidden layers is directly proportional to MAD as well. Hence for all cases one hidden layer is selected.

12. During training of the model, local minima and global minima are obtained by 5400000 and 7500000 epochs respectively. However, models have produced acceptable results using 200 epochs as well.

13. In deterministic forecast, preceding n years meteorological parameter time series have been used as input to the model to observed $(n+1)^{th}$ year data as target.

14. In parametric forecast, n number of parameters or predictors (i.e., meteorological data which is physically connected with target meteorological data) data time series of each year are used as input to the model to predict target meteorological data.

15. The deterministic neural network technique is capable to explain the non-linearity of the long-range rainfall variable data time series of smaller region. However, results show very unfortunate performance in independent period.

16. Identification of predictors by observing their physical linkage and degree of relationship with long-range TMRF is very difficult. However, eight meteorological parameters as predictors have been identified for Ambikapur region of Surguja district, India.

17. Thus 8-parameter BPN model has been developed to recognize pattern and prediction of TMRF over this region. Result shows very good performance as compared to deterministic model. If we are capable to indentify appropriate predictors for an Indian geographical region then this model is extremely applicable.

18. In this study, we have applied 8 parameters only. However, they may correlate to each other. Thus PCA have been done to reduce correlated parameters. And obtained 02 PCs. This exercise has

provided two fixations. First it reduces unnecessary input variables. And second, it reduces unknown variables (i.e., weights) from the model. as a result, we reduced total training time.

19. PCs-BPN model has developed. It is found that this model also capable to explain internal dynamics of monsoon rainfall over the district place and similar performance as parametric model has been shown.

20. Finally, a hybrid BPN model is introduced, wherein output of all three models has been used as input of a new BPN model to observed TMRF as an output. Actual benefit of this architecture is that, we may able to more trainable weights in the network to develop relationship between dependent and independent variables. In the parametric model, we have seen that break-down situations. It means that we cannot introduce more trainable weights. In this model 40 weights from deterministic forecast, 31 weights from parametric forecast, 9 trainable weights from PC based parametric forest is used to develop relationship. Two neurons in hidden layer, three input vectors in input layer are used in new BPN. Thus total of 11 trainable weights including biases plus 40, 31, 9 trained weights have been used in this hybrid model.

21. It is concluded that, hybridization of parametric and deterministic model in this application may be useful by utilize different set of input data in identification of internal dynamics of TMRF over the district however; not a better solution can declare especially in this case.

22. Moreover, long-range monsoon rainfall forecasting system over the district level has a longer lead-time as they can be made a year in

advance. Thus the system developed is an innovative one to meet the objectives.

Future Scope

In this study, it has been found that variability of Indian district level or very small-scale geographical region monsoon rainfall is due to external forcing, i.e., global weather parameter and also the internal variability within the time series itself. If external forcing is assumed to be same then also there remains variability within the time series that can only be explained if one can able to predict the internal dynamical behavior of the weather data time series. The pattern recognition and prediction in a deterministic or parametric approach through ANN technique based on back-propagation algorithm has been found to be a most efficient way by which internal dynamics of rainfall time series data can be successfully identified. Moreover, pattern recognition and prediction have a longer lead-time as they can be made a year in advance. No other model (except the neural network model) so far has been able to forecast district level long-range weather parameter so accurately.

Back-propagation using gradient descent (algorithm 2.1 and 2.2) often converges very slowly or not at all. On large-scale problems its success depends on user-specified learning rate and momentum parameters. There is no automatic way to select these parameters, and if incorrect values are specified the convergence may be exceedingly slow, or it may not converge at all. While back-propagation with gradient descent still is used in the model development, it is no longer considered to be the best or fastest algorithm. Instead of gradient descent, conjugate gradient algorithm can also be used as a future work to adjust weight values using the gradient during the backward propagation of errors through the network. Compared to gradient descent, the conjugate gradient algorithm takes a more direct path to the optimal set of weight values. Usually, conjugate gradient is significantly faster and more robust than gradient descent. Conjugate gradient also does not require the user to specify learning rate and momentum parameters. In addition, performance of the RBF, PNN, GRNN, and CPN network can also be compared with the applied BPN model as a future work.

Limitations

The limitations of the ANN models are as listed:

1. The number of learning steps may be high, and also the learning phase has exhaustive calculation.

2. The selection of the number of input vectors and hidden neurons in the network is really a problem.

3. Time required for the training is very high as discussed in Section 2.9.

4. Back-propagation using gradient descent (algorithm 2.1 and 2.2) often converges very slowly or not at all.

5. The training may some-time cause "nervousness" due to program execution hanged up (heap space problem) as shown in following Fig.

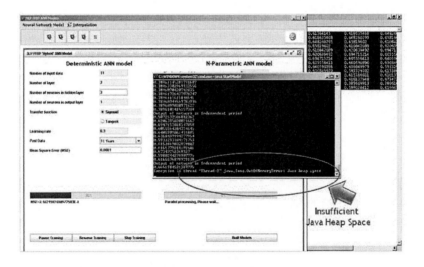

Fig. Training Program Execution Hanged up due to "Exception in Thread Java.lang.Out.Of.MemoryError: Java Heap Space". While, allocated Heap Space is Minimum 512 MB and Maximum 1024 MB.

Publications

1. National/International Journals

[1]. Karmakar, S., Kowar, M. K., Guhathakurta, P., *Evolving and Evaluation of 3LP FFBP Deterministic ANN Model for District Level Long Range Monsoon Rainfall Prediction*, J. Environmental Science & Engineering, , National Environmental Engineering Research Institute, Nagpur, INDIA, ISSN 0367-827, Vol 51, No. 2, pp. 137-144, April 2009.

URL: http://www.neeri.res.in/jese.html.

[2]. Karmakar, S., Kowar, M. K., Guhathakurta, P., *Long Range Summer Monsoon Rainfall Forecasting over Subdivision Chhattisgarh by Deterministic Neural Network*, CSVTU Research Journal, Chhattisgarh Swami Vivekanand Technical university, Bhilai, INDIA, ISSN 0974-8725, Vol. 02, No. 01, pp. 37-41, January 2009.

URL:http://www.csvtu.ac.in/pdf_doc/abstractofcsvturesearchgernal2009.pdf

[3]. Karmakar, S., Kowar, M. K., Guhathakurta, P., *Development of ANN Models for Temperature parameters pattern recognition over the smaller scale geographical region- District*, International J. of Engg. Research & Indu. Appls. (IJERIA), Acent Publication Pune, INDIA ISSN 0974-1518, Vol. 1, No. IV, pp. 111-121, 2009.

URL: http://www.ascent-journals.com/ijeria_contents_Vol1No6.htm

[4]. Karmakar, S., Kowar, M. K., Guhathakurta, P., *Artificial Neural Network Skeleton in Deterministic Forecast to Recognize Pattern of*

TMRF Ambikapur District, CSVTU Research Journal, Chhattisgarh Swami Vivekanand Technical university, Bhilai, INDIA, ISSN 0974-8725, Vol. 02, No. 02, pp. 41-45, January 2009.

[5]. Karmakar, S., Kowar, M. K., Guhathakurta, P., *Development of a 6 Parameter Artificial Neural Network Model for Long-Range July Rainfall Pattern Recognition over the Smaller Scale Geographical Region –District*, CSVTU Research Journal, Chhattisgarh Swami Vivekanand Technical university, Bhilai, INDIA, ISSN 0974-8725, Vol. 03, No. 01, pp.36-40, January 2010.

[6]. Shrivastava, G., Karmakar, S., Kowar, M.K., Guhathakurta, P., Application of Artificial Neural Networks in Weather Forecasting: A Comprehensive Literature Review, International Journal of Computer Applications, Vol. 51, No.18, Foundation of Computer Science, New York, USA, pp. 17-29, August 2012.

[7]. Varghese, B., Shrivastava, G., Karmakar, S., Kowar, M.K., Guhathakurta, P., Architecture of Artificial Neural Network in Identification of Internal Dynamics and Prediction of Dynamic System Rainfall Data Time Series, International Journal of Computer Science and Informatics (IJCSI), Vol.1. No. 4, pp. 1-9.

2. International/National Conferences

[1] Karmakar, S., Kowar, M. K., Guhathakurta, P., *Development of an 8-Parameter Probabilistic Artificial Neural Network Model for Long-Range*

Monsoon Rainfall Pattern Recognition over the Smaller Scale Geographical Region –District, IEEE Computer Society, *Proc.* of International Conference on Emerging trends in Engineering and Technology, Washington, DC, USA, ISBN 978-0-7695-3267-7, pp. 569-574, July 2008.

URL:

http://www2.computer.org/portal/web/csdl/doi/10.1109/ICETET.20
08.225

[2]   Karmakar, S., Kowar, M. K., Guhathakurta, P., *Development of Artificial Neural Network Models for Long-Range Meteorological Parameters Pattern Recognition over the Smaller Scale Geographical Region-District*, *Proc.* of IEEE International Conference on Industrial and Information system, IEEE Computer Society, Washington, DC, USA, ISBN 978-1-4244-2806-9, pp. 1-6, Dec. 2008.

URL:http://ieeexplore.ieee.org/xpl/freeabs_all.jsp?isnumber=4798312
&arnumber=4798370&count=184&index=54.

[3]   Karmakar, S., Kowar, M. K., Guhathakurta, P., *Long-Range Monsoon Rainfall Pattern Recognition & Prediction for the Subdivision 'EPMB' Chhattisgarh Using Deterministic & Probabilistic Neural Network*, *Proc.* of IEEE International Conference on Advances in Pattern Recognition, IEEE Computer Society, Washington, DC, USA, ISBN 978-0-7695-3520-3, pp. 376-370, Feb. 2009. URL:

http://www2.computer.org/portal/web/csdl/doi/10.1109/ICAPR.200 9.24.

[4]   Karmakar, S., Kowar, M. K., Guhathakurta, P., *Spatial Interpolation of Rainfall Variables using Artificial Neural Network, Proc.* of ACM International Conference on Advancement of Computing, Communication and Control, ACM New York, NY, USA, ISBN ACM 978-1-60558-351-8, pp. 547-552, 2009.URL:

http://portal.acm.org/citation.cfm?id=1523212&dl=GUIDE&coll=G UIDE.

[5].   Deterministic FFBP ANN Design to Recognize Internal Dynamics of Chaotic Motion , Proc. of 1st International Conference on Emerging Trends in Soft Computing & ICT, Guru Ghasidas Central University, Bhilaspur, Chhattisgarh, India, 2011.

[6].   Identification of Internal Dynamics of Monsoon Rainfall (Chaos), Technologia 2010, MPCCET, Bhilai, 24th-25th Feb 2010.

[7].   Applications of ANN to identify Motion of Chaos, AICTE National Seminar on Modern Trend in Applied Mathematics, Bhilai Institute of Technology, Durg, 8th-9th Jan. 2010.

3.   Book

[1].   Karmakar, S., Kowar, M. K., Guhathakurta, P., Applications of Neural Network in Weather Forecasting, ISSN: 978-3-659-26670-

6, LAP LAMBERT Academic publications, Germany, Saarbruchen 2012.

References

[1]. Walker, G. T., 1923, "Correlation in Seasonal Variations of Weather, III. A Preliminary Study of World Weather". Mem. India Meteorol. Dep., XXIV, 75–131.

[2]. Walker, G. T., 1924, "Correlation in Seasonal Variations of Weather, IV. A Further Study of World Weather". Mem. India Meteorol. Dep., XXIV, 275– 332.

[3]. Gowariker, V., Thapliyal, V., Sarker, R. P., Mandal, G. S. and Sikka, D. R., 1989, "Parametric and Power Regression Models: New Approach to Long Range Forecasting of Monsoon Rainfall in India". Mausam, 40, 115– 122.

[4]. Gowariker, V., Thapliyal, V., Kulshrestha, S. M., Mandal, G. S., Sen Roy, N., and Sikka, D. R., 1991, "A Power Regression Model for Long Range Forecast of Southwest Monsoon Rainfall over India". Mausam, 42, 125–130.

[5]. Thapliyal, V., and Kulshrestha, S. M., 1992, "Recent Models for Long Range Forecasting of Southwest Monsoon Rainfall over India". Mausam, 43, 239–248.

[6]. Thapliyal, V., 1997, "Preliminary and Final Long Range Forecasts for Seasonal Monsoon Rainfall over India". J. Arid Environ., 36, 385–403.

[7]. Rajeevan, M., Guhathakurta, P. and Thapliyal, V., 2000, "New Models for Long Range Forecasting of Monsoon Rainfall over Northwest and Peninsular India". Meteorol. Atmos. Phys., 73, 211–225.

[8].   Rajeevan M., 2001, "Prediction of Indian Summer Monsoon: Status, Problems and Prospects". Current Science, 81, 1451–1457.

[9].   Thapliyal, V., and Rajeevan, M., 2003, "Updated Operational Models for Long-Range Forecasts of Indian Summer Monsoon Rainfall". Mausam, 54, 495–504.

[10].   Rajeevan, M., Pai, D.S., Dikshit, S.K., and Kelkar, R. R., 2004, "IMD's New Operational Models for Long-Range Forecast of Southwest Monsoon Rainfall over India and Their Verification for 2003", Current Science, 86 (3), 422-431.

[11].   Guhathakurta, P., 2000, "New Models for Long Range Forecasts of Summer Monsoon Rainfall over North West and Peninsular India", Meteor. & Atomos. Phys.,73 (3), 211-255.

[12].   Guhathakurta, P., Rajeevan, M., and Thapliyal, V., 1999, "Long Range Forecasting Indian Summer Monsoon Rainfall by Hybrid Principal Component Neural Network Model", Meteorology and Atmospheric Physics, Springer-Verlag, Austria.

[13].   Parthasarathy, B., Rupa Kumar, K., and Munot, A. A., 1991, "Evidence of Secular Variations in Indian Summer Monsoon Rainfall Circulation Relationships". J. Climate, 4, 927–938.

[14].   Hastenrath, S., and Greischar, L., 1993, "Changing Predictability of Indian Monsoon Rainfall Anomalies". Proc. Indian Acad. Sci. (Earth Planet. Sci.), 102, 35–47.

[15].   Guhathakurta P., (2006), "Long-Range Monsoon Rainfall Prediction of 2005 for the Districts and Sub-Division Kerala with Artificial Neural Network", Current Science, 90(6), pp-773-779.

197

[16]. Krishnamurthy, V., and Kinter, J. L., 2002, "The Indian Monsoon and its Relation to Global Climate Variability". Global Climate – Current Research and Uncertainties in the Climate System (eds Rodo, X. and Comin, F. A.), 186–236.

[17]. Krishnamurthy, V., and Kirtman, B. P., 2003, "Variability of the Indian Ocean: Relation to Monsoon and ENSO". Q. J. R." Meteorol. Soc., 129, 1623–1646

[18]. Sahai, A. K., Grimm, A. M., Satyan. V. and Pant, G. B., 2002, "Prospects of Prediction of Indian Summer Monsoon Rainfall using Global SST Anomalies". IITM Research Report No. RR-093.

[19]. Guhathakurta, P., 1999, "Long Range Forecasting Indian Summer Monsoon Rainfall By Principle Component Neural Network Model", Meteor. & Atomos. Phys., 71, 255-266.

[20]. Guhathakurta, P., 1998, "A Hybrid Neural Network Model for Long Range Prediction of All India Summer Monsoon Rainfall", Proceedings of WMO international workshop on dynamical extended range forecasting, Toulouse, France, November 17-21, 1997, PWPR No. 11, WMO/TD. 881, pp 157-161.

[21]. Basu Sujit, Andharia H I, 1992, "The Chaotic time series of Indian Monsoon rainfall and its prediction", Proc. Indian Acad. Sci., Vol. 101, No. 1, pp 27-34.

[22]. Chow T. W. S. and Cho S. Y., 1997, "Development of a Recurrent Sigma-Pi Neural Network Rainfall Forecasting System in Hong Kong", Springer-Verlag, pp. 66-75.

[23]. Lee Sunyoung, Cho Sungzoon, Wong Patrick M., 1998, "Rainfall Prediction Using Artificial Neural Networks", Journal of Geographic Information and Decision Analysis, vol. 2, no. 2, pp. 233 – 242.

[24]. Hsieh, W.H., and Tang, B., 1998, "Applying Neural Network Models to Prediction and Analysis in Meteorology and Oceanography". Bull. Amer. Met. Soc., 79, 855-1870.

[25]. Dawson Christian W., Wilby Robert, "An artificial neural network approach to rainfall runoff modeling", 1998, Hydrological Sciences—Journal, 43(1), pp. 47-66.

[26]. Guhathakurta1 P., Rajeevan2 M., and Thapliyal2 V., 1999, "Long Range Forecasting Indian Summer Monsoon Rainfall by a Hybrid Principal Component Neural Network Model", Meteorol. Atmos. Phys., 71, pp. 255-266.

[27]. Ricardo, M., Trigo, Jean, P., Palutikof, 1999, "Simulation of Daily Temperatures for Climate Change Scenarios over Portugal: A Neural Network Model Approach", University of East Anglia, Norwich, NR4 7TJ, United Kingdom, climate research (Clim., Res.), 13, 45–59, 1999.

[28]. Jones, C., Peterson, P., 1999, "A New Method for Deriving Ocean Surface Specific Humidity and Air Temperature: An Artificial Neural Network Approach", J. Applied Meteorology, American Meteorological Society, 38, 1229-1245.

[29]. Guhathakurta, P., 1999, "A Short Term Prediction Model for Surface Ozone At Pune: Neural Network Approach", Vayu

mandal, Special issue on Asian monsoon and pollution over the monsoon environment, 29(1-4), 355-358.

[30]. Guhathakurta, P., 1999, "A Neural Network Model for Short Term Prediction of Surface Ozone at Pune", Mausam, 50(1), 91-98.

[31]. Toth E. *, Brath A., Montanari A., 2000, "Comparison of short-term rainfall prediction models for real-time flood forecasting", Journal of Hydrology, Elsevier, 239, pp. 132-147.

[32]. Luk Kin C, Ball J. E. and Sharma A., 2001, "An Application of Artificial Neural Networks for Rainfall Forecasting", 33, pp. 883-699.

[33]. Michaelides Silas Chr,*, Pattichis Constantinos S. and Kleovouloub Georgia, 2001, "Classification of rainfall variability by using artificial neural networks", International Journal Of Climatology, pp. 1401–1414.

[34]. Chang Fi-John, Liang Jin-Ming, and Chen Yen-Chang,2001, "Flood Forecasting Using Radial Basis Function Neural Networks", IEEE Transactions On Systems, Man, And Cybernetics, vol. 31, no. 4, pp. 530-535.

[35]. Brath A., Montanari A. and Toth E., 2001, "Neural networks and non-parametric methods for improving ealtime flood forecasting through conceptual hydrological models", Hydrology and Earth System Sciences, 6(4), pp. 627-640.

[36].   Rajurkar M. P., Kothyari U. C., Chaube U. C., 2002, "Artificial neural networks for daily rainfall-runoff modelling", Hydrologkal Sciences-Journals, 47(6), pp. 865-877.

[37].   Harun Sobri, Nor Irwan Ahmat & Mohd. Kassim Amir Hashim, 2002, "Artificial neural network model for rainfall-runoff relationship", Journal Technology, Malaysia, pp. 1-12.

[38].   Iseri, Y. G. C., Dandy, R., Maier, A., Kawamura, and Jinno, K., 2002, "Medium Term Forecasting of Rainfall Using Artificial Neural Networks", Part 1 background and methodology, Journal of Hydrology , 301 (1-4), 1834-1840.

[39].   Silva, A. P., 2003, "Neural Networks Application to Spatial Interpolation of Climate Variables", Carried Out By, STSM on the Framework of COST 719 ZAMG, Vienna 6-10 October 2003.

[40].   Snell., Seth, E., Gopal, Sucharita, Kaufmann, Robert, K., 2003, "Spatial Interpolation of Surface Air Temperatures Using Artificial Neural Networks: Evaluating Their Use for Downscaling GCMs", Journal of Climate, 13 (5), 886-895.

[41].   Blender, R., 2003, "Predictability Study of the Observed and Simulated European Climate using Linear Regression", Q. J. R. Meteorol. Soc. (2003), 129, 2299–2313.

[42].   Maqsood Imran, Khan Muhammad Riaz, Abraham Ajith, 2004, Neural Comput & Applic, 13, pp. 112–122.

[43].   Pasero Eros, Moniaci Walter, 2004, "Artificial Neural Networks for Meteorological Nowcast", IEEE international Conference,

[44]. Lekkas D.F.,* Onofl C., Lee1 M. J., Baltas2 E.A., 2004, "Application of artificial neural networks for flood forecasting", Global Nest, Vol 6, No 3, pp 205-211.

[45]. Shu Chang and Burn Donald H., 2004, "Artificial neural network ensembles and their application in pooled flood frequency analysis", Water Resources Research, vol. 40.

[46]. Nayaka P.C.,*, Sudheerb K.P.,1Ranganc, D.M.,2, Ramasastrid K.S.,3, 2004, "A neuro-fuzzy computing technique for modeling hydrological time series", Journal of Hydrology, Elsevier, 291, pp. 52–66.

[47]. Wu Jy S., P.E., ASCE1 M.; Han2 Jun; Annambhotla3 Shastri; and Scott Bryant4, 2004, "Artificial Neural Networks for Forecasting Watershed Runoff and Stream Flows", Journal Of Hydrologic Engineering, 216.

[48]. Abdel-Aal R.E. *, 2004, "Hourly temperature forecasting using abductive networks", Elsevier, 17, pp. 543–556.

[49]. Lee Tsong-Lin, 2004, "Back-propagation neural network for long-term tidal predictions", Elsevier, 31, pp. 225–238.

[50]. Chaudhuri Sutapa, Chattopadhyay Surajit, 2005, "Neuro-computing based short range prediction of some meteorological parameters during the pre-monsoon season", Springer-Verlag, 9, pp. 349–354.

[51]. Lin Gwo-Fong * and Chen Lu-Hsien, 2005, "Application of an artificial neural network to typhoon rainfall forecasting", Hydrological Processes, 19, pp. 1825–1837.

[52]. Vandegriff Jon *, Wagstaff Kiri, George Ho, Plauger Janice, 2005, "Forecasting space weather: Predicting interplanetary shocks using neural networks", Elsevier, vol 36, pp. 2323–2327.

[53]. KISI Ozgur, 2005, "Daily River Flow Forecasting Using Artificial Neural Networks and Auto-Regressive Models", Turkish J. Eng. Env. Sci., vol 29, pp. 9 - 20.

[54]. Maqsood Imran, Khan Muhammad Riaz, Huang Guo H., Abdalla Rifaat, 2005, "Application of soft computing models to hourly weather analysis in southern Saskatchewan, Canada", Elsevier, vol 18, pp. 115–125.

[55]. Somvanshi V.K., Pandey O.P., Agrawal P.K., Kalanker1 N.V., Prakash M.Ravi and Chand Ramesh, 2006, "Modelling and prediction of rainfall using artificial neural network and ARIMA techniques", J. Ind. Geophys. Union, Vol.10, No.2, pp.141-151.

[56]. Srikalra Niravesh and Tanprasert Chularat, 2006, "Rainfall Prediction for Chao Phraya River using Neural Networks with Online Data Collection", Malaysia, pp. 13-15.

[57]. Kumarasiri A.D.and Sonnadara D.U.J., 2006, "Rainfall Forecasting: An Artificial Neural Network Approach", Proceedings of the Technical Sessions, vol 22, pp. 1-13.

[58]. Kumar D. Nagesh, Reddy M. Janga and Maity Rajib, 2006, "Regional Rainfall Forecasting using Large Scale Climate Teleconnections and Artificial Intelligence Techniques", Journal of Intelligent Systems, Vol. 16, No.4, pp. 307-322.

[59]. Guhathakurta, P., 2006, "Long-Range Monsoon Rainfall Prediction of 2005 for the Districts and Sub-Division Kerala With Artificial Neural Network", Current Science, 90 (6), pp-773-779.

[60]. Bustami Rosmina, I BessaihNabi, Charles Bong, Suhaila Suhaili, 2007, "Artificial Neural Network for Precipitation and Water Level Predictions of Bedup River", International Journal of Computer Science, vol 34:2.

[61]. Paras, Mathur Sanjay, Kumar Avinash, and Chandra Mahesh, 2007, "A Feature Based Neural Network Model for Weather Forecasting", World Academy of Science, Engineering and Technology, vol 34, pp. 66-73.

[62]. Hayati Mohsen, and Mohebi Zahra, 2007, "Application of Artificial Neural Networks for Temperature Forecasting", World Academy of Science, Engineering and Technology, vol 28, pp. 275-279.

[63]. Morid Saeid, Smakhtin Vladimir and Bagherzadeh K., 2007, "Drought forecasting using artificial neural networks and time series of drought indices", Royal Meteorological Society, vol. 27, pp. 2103–2111.

[64]. Hayati Mohsen, and Shirvany Yazdan, 2007, "Artificial Neural Network Approach for Short Term Load Forecasting for Illam Region", World Academy of Science, Engineering and Technology, vol. 28, pp. 280-284.

[65]. Lucio P. S., Conde F. C., Cavalcanti I. F. A., Serrano A. I., Ramos A. M., and Cardoso A. O., 2007, "Spatiotemporal monthly rainfall

reconstruction via artificial neural network – case study: south of Brazil", Advances in Geosciences, vol. 10, pp. 67–76.

[66]. Hartmann, Heikea * Becker Stefanb and Kinga Lorenz, 2007, "Predicting summer rainfall in the Yangtze River basin with neural networks", Royal Meteorological Society.

[67]. Aliev R. A., Fazlollahi B., Aliev R. R., Guirimov B., 2008, "Linguistic time series forecasting using fuzzy recurrent neural network", Soft Comput, vol. 12, pp. 183–190.

[68]. Chattopadhyay Surajit and Chattopadhyay Goutami, 2008, "Identification of the best hidden layer size for three layered neural net in predicting monsoon rainfall in India", Journal of Hydroinformatics, vol. 10(2), pp. 181-188.

[69]. Hung N. Q., Babel M. S., Weesakul S., and Tripathi N. K., 2008, "An artificial neural network model for rainfall forecasting in Bangkok, Thailand", Hydrology and Earth System Sciences, vol. 5, pp. 183–218.

[70]. Aytek Ali, Asce M and Alp Murat, 2008, "An application of artificial intelligence for rainfall–runoff modeling", J. Earth Syst. Sci., vol. 117, No. 2, pp. 145–155.

[71]. Chattopadhyay Goutami, Chattopadhyay Surajit, Jain Rajni, 2008, "Multivariate forecast of winter monsoon rainfall in India using SST anomaly as a predictor: Neurocomputing and statistical approaches",

[72]. Win Khaing Mar and Thu Naing Thinn, 2008, "Optimum Neural Network Architecture for Precipitation Prediction of

Myanmar", World Academy of Science, Engineering and Technology, vol. 48, pp. 130-134.

[73]. Karmakar, S, Kowar, M.K., Guhathakurta, P., Development of Artificial Neural Network Models for Long-Range Meteorological Parameters Pattern Recognition over the Smaller Scale Geographical Region-District, IEEE Xplore 2.0, IEEE Computer Society, Washington, DC, USA, ISBN 978-1-4244-2806-9, pp. 1-6, Dec. 2008. URL:http://ieeexplore.ieee.org/xpl/freeabs_all.jsp?isnumber=47983 12&arnumber=4798370&count=184&index=54.

[74]. Karmakar, S, Kowar, M.K., Guhathakurta, P., Evolving and Evaluation of 3LP FFBP Deterministic ANN Model for District Level Long Range Monsoon Rainfall Prediction, J. Environmental Science & Engineering, , National Environmental Engineering Research Institute, Nagpur, INDIA, ISSN 0367-827, Vol 51, No. 2, pp. 137-144, April 2009. URL: http://www.neeri.res.in/jese.html.

[75]. Hocaoglu Fatih O., Oysal Yusuf, Kurban Mehmet, 2009, "Missing wind data forecasting with adaptive neuro-fuzzy inference system", Springer-Verlag London, vol. 18, pp. 207–212.

[76]. Solaimani Karim, 2009, "Rainfall-runoff Prediction Based on Artificial Neural Network (A Case Study: Jarahi Watershed)", American-Eurasian J. Agric. & Environ. Sci., vol. 5(6), pp. 856-865.

[77]. KOŠCAK Juraj, JAKŠA Rudolf., SEPEŠI Rudolf, SINCÁK Peter., 2009, "Weather forecast using Neural Networks", 9th Scientific Conference of Young Researchers

[78]. Karamouz M.; . Fallahi M, Nazif S. and Farahani M. Rahimi, 2009, "Long Lead Rainfall Prediction Using Statistical Downscaling and Arti cial Neural Network Modeling", Transaction A: Civil Engineering, Vol. 16, No. 2, pp. 165-172.

[79]. Widjanarko Bambang Otok, Suhartono, 2009, "Development of Rainfall Forecasting Model in Indonesia by using ASTAR, Transfer Function, and ARIMA Methods", European Journal of Scientific Research, Vol.38 No.3, pp.386-395.

[80]. Nekoukar Vahab, Taghi Mohammad, Beheshti Hamidi, 2010, "A local linear radial basis function neural network for financial time-series forecasting", Springer Science, vol. 23, pp. 352–356.

[81]. Weerasinghe H.D.P.,. Premaratne H.L and Sonnadara D.U.J., 2010, "Performance of neural networks in forecasting daily precipitation using multiple sources", J.Natn.Sci.Foundation Sri Lanka, vol. 38(3), pp. 163-170.

[82]. Luenam Pramote, Ingsriswang Supawadee, Ingsrisawang Lily, Aungsuratana Prasert, and Khantiyanan Warawut, 2010, "A Neuro-Fuzzy Approach for Daily Rainfall Prediction over the Central Region of Thailand", ISSN 2010, vol. 1.

[83]. Wu C. L.,. Chau1 K. W, and Fan C., 2010, "Prediction of Rainfall Time Series Using Modular Artificial Neural Networks

Coupled with Data Preprocessing Techniques", Journal of Hydrology, Vol. 389, No. 1-2, pp. 146-167.

[84].   Nastos Panagiotis, Moustris Kostas, Larissi Ioanna, and Paliatsos Athanasios, 2010, "Rain intensity forecast using Artificial Neural Networks in Athens, Greece", Geophysical Research Abstracts, Vol. 12.

[85].   Patil C. Y. and Ghatol A. A., 2010, "Rainfall forecasting using local parameters over a meteorological station: an artificial neural network approach", International J. of Engg. Research & Indu. Appls, Vol.3, No. II, pp 341-356.

[86].   Tiron Gina, and Gosav Steluța, 2010, "The july 2008 rainfall estimation from BARNOVA WSR-98 D Radar using artificial neural network", Romanian Reports in Physics, Vol. 62, No. 2, pp. 405–413.

[87].   Goyal Manish Kumar, Ojha Chandra Shekhar Prasad, 2010, "Analysis of Mean Monthly Rainfall Runoff Data of Indian Catchments Using Dimensionless Variables by Neural Network", Journal of Environmental Protection, vol. 1, pp. 155-171.

[88].   .Vamsidhar Enireddy Varma K.V.S.R.P..Sankara Rao P satapati Ravikanth, 2010, "Prediction of Rainfall Using Backpropagation Neural Network Model", International Journal on Computer Science and Engineering, Vol. 02, No. 04, pp. 1119-1121.

[89].   Haghizadeh Ali, Teang shui Lee, Goudarzi Ehsan, 2010, "Estimation of Yield Sediment Using Artificial Neural Network at

Basin Scale", Australian Journal of Basic and Applied Sciences, vol. 4(7), pp. 1668-1675.

[90]. Subhajini A . C. and Joseph Raj V., 2010, "Computational Analysis of Optical Neural Network Models to Weather Forecasting", International Journal of Computer Science Issue, Vol.7, Issue 5, pp. 327-330.

[91]. Omer Faruk Durdu, 2010, "A hybrid neural network and ARIMA model for water quality time series prediction.", Elsevier, vol 23, pp. 586–594

[92]. . Soman Saurabh S, Zareipour Hamidreza, Malik Om, and Mandal Paras, 2010, "A Review of Wind Power and Wind Speed Forecasting Methods With Different Time Horizons".

[93]. Khalili Najmeh, Khodashenas Saeed Reza, Davary Kamran and Karimaldini Fatemeh, 2011, "Daily Rainfall Forecasting for Mashhad Synoptic Station using Artificial Neural Networks", International Conference on Environmental and Computer Science, vol.19, pp. 118-123.

[94]. Pan Tsung-Yi, Yang Yi-Ting, Kuo Hung-Chi, Tan Yih-Chi, Lai1 Jihn-Sung, Chang Tsang-Jung, Lee Cheng-Shang, and Hsu Kathryn Hua, 2011, "Improvement of Statistical Typhoon Rainfall Forecasting with ANN-Based Southwest Monsoon Enhancement", Terr. Atmos. Ocean. Sci, Vol. 22, No. 6, pp. 633-645.

[95]. Joshi Jignesh, Patel Vinod M., 2011, "Rainfall-Runoff Modeling Using Artificial Neural Network (A Literature Review)", National Conference on Recent Trends in Engineering & Technology.

[96]. El-Shafie Amr H., Shafie A. El-, Mazoghi Hasan G. El, Shehata A. and Taha Mohd. R., 2011, "Artificial neural network technique for rainfall forecasting applied to Alexandria, Egypt",

[97]. Mekanik F., Lee T.S. and Imteaz M. A., 2011, "Rainfall modeling using Artificial Neural Network for a mountainous region in West Iran".

[98]. Kaur Amanpreet, Singh Harpreet, 2011, "Artificial Neural Networks in Forecasting Minimum Temperature", International Journal of Electronics & Communication Technology, Vol. 2, Issue 3, pp. 101-105.

[99]. El-Shafie A., Jaafer O. and Akrami Seyed Ahmad, 2011, "Adaptive neuro-fuzzy inference system based model for rainfall forecasting in Klang River, Malaysia", International Journal of the Physical Sciences, Vol. 6(12), pp. 2875-2888.

[100]. Tripathy Asis Kumar, Mohapatra Suvendu, Beura Shradhananda, Pradhan Gunanidhi, 2011. "Weather Forecasting using ANN and PSO", International Journal of Scientific & Engineering Research, Volume 2, Issue 7, pp.1-5.

[101]. El-Shafie A, Noureldin A., Taha M. R., and Hussain A., 2011, "Dynamic versus static neural network model for rainfall forecasting at Klang River Basin, Malaysia", Hydrol. Earth Syst. Sci., vol. 8, pp. 6489–6532.

[102]. Geetha G.,. Selvaraj R Samuel, 2011, "Prediction of monthly rainfall in Chennai using back propagation neural network model", International Journal of Engineering Science and Technology, Vol. 3 No. 1, pp. 211-213.

[103]. Reshma T., Reddy K. Venkata, Pratap Deva, 2011, "Determination of Distributed Rainfall- Runoff Model Parameters Using Artificial Neural Network", International Journal of Earth Sciences and Engineering, Volume 04, No 06, pp. 222-224

[104]. Raju M.Mohan,. Srivastava R. K, Bisht Dinesh C. S., Sharma H. C., and Kumar Anil, 2011, "Development of Artificial Neural-Network-BasedModels for the Simulation of Spring Discharge", Hindawi Publishing Corporation, Volume 2011, pp. 1-11.

[105]. Kavitha M.Mayilvaganan,.Naidu K.B, 2011, "ANN and Fuzzy Logic Models for the Prediction of groundwater level of a watershed", International Journal on Computer Science and Engineering, Vol. 3 No. 6, pp. 2523-2530.

[106]. El-shafie A., Mukhlisin M., Najah Ali A. and Taha M. R., 2011, "Performance of artificial neural network and regression techniques for rainfall-runoff prediction", International Journal of the Physical Sciences, Vol. 6(8), pp. 1997-2003.

[107]. Afshin Sarah, Fahmi Hedayat, Alizadeh Amin, Sedghi Hussein and Kaveh Fereidoon, 2011, "Long term rainfall forecasting by integrated artificial neural network-fuzzy logic-wavelet model in Karoon basin", Scientific Research and Essays, Vol. 6(6), pp. 1200-1208.

[108]. Siou1 Line Kong A, Johannet Anne, Borrell Valérie, Pistre Séverin, 2011, "Complexity selection of a neural network model for karst flood forecasting: The case of the Lez Basin (southern France)", Journal of Hydrology, vol. 403, pp. 367-380.

[109]. Saima H., Jaafar J., Belhaouari S., Jillani T.A., 2011, "Intelligent Methods for Weather Forecasting: A Review", IEEE,

[110]. Sawaitul Sanjay D., Prof. Wagh K. P., Dr. Chatur P. N., 2012, "Classification and Prediction of Future Weather by using Back Propagation Algorithm-An Approach", International Journal of Emerging Technology and Advanced Engineering, Volume 2, Issue 1, pp. 110-113.

[111] Karmakar, S, Kowar, M.K., Guhathakurta, P., Development of an 8-Parameter Probabilistic Artificial Neural Network Model for Long-Range Monsoon Rainfall Pattern Recognition over the Smaller Scale Geographical Region –District, IEEE Computer Society, IEEE Xplore 2.0,, DC, USA, ISBN 978-0-7695-3267-7, pp. 569-574, July 2008. http://www2.computer.org/portal/web/csdl/doi/10.1109/ICETET .2008.225. http://portal.acm.org/citation.cfm?id=1445475

[112]. Karmakar, S, Kowar, M.K., Guhathakurta, P., Long-Range Monsoon Rainfall Pattern Recognition & Prediction for the Subdivision 'EPMB' Chhattisgarh Using Deterministic & Probabilistic Neural Network, IEEE Xplore 2.0, IEEE Computer Society, Washington, DC, USA, ISBN 978-0-7695-3520-3, pp. 376-370, Feb. 2009.

http://www2.computer.org/portal/web/csdl/doi/10.1109/ICAPR.2009.24.

[113]. NBSSLUP., 2006, "Soil Series of Chhattisgarh", Department of Agriculture, Government of Chhattisgarh NBSS Publ., 85. Nagpur, Maharastra, India.

[114]. Bryan, B.A., Adams, J.M., 2001 "Quantitative and Qualitative Assessment of the Accuracy of Neurointerpolated Annual Mean Precipitation and Temperature Surfaces for China", Cartography, 30 (2), Perth, Australia.

[115]. Basu, S., Andhariya, H.I., 1991, "The Chaotic Time Series of Indian Rainfall and Its Prediction", Proc. Ind. Acad. Sci., 101. 27-34.

[116]. Rajeevan, M., Pai, D. S., and Thapliyal, V., 1998, "Spatial and Temporal Relationships Between Global and Surface Air Temperature Anomalies and Indian Summer Monsoon". Meteorol. Atmos. Phys.,66, 157–171.

[117]. Rajeevan, M., 2002, "Winter Surface Pressure Anomalies Over Eurasia and Indian Summer Monsoon", Geophys. Res. Lett., 29, 94.1-94.4.

[118]. Rajeevan, M., Pai, D. S., and Thapliyal, V., 2002, "Predictive Relationships Between Indian Ocean Sea Surface Temperatures and Indian Summer Monsoon Rainfall". Mausam, 53, 337–348.

[119]. Hu, M.J.C., 1964. Application of the adaline system to weather forecasting. Master Thesis, Technical Report 6775-1, Stanford Electronic Laboratories, Stanford, CA, June.

[120]. Rumelhart, D.E., Hinton, G.E., Williams, R.J., 1986. Learning representations by backpropagating errors. Nature 323 (6188), 533–536.

[121]. Rumelhart, D.E., Hinton, G.E., Williams, R.J., 1986. Learning internal representation by back- propagating errors. In: Rumelhart, D.E., McCleland, J.L., the PDP Research Group (Eds.), Parallel Distributed Processing: Explorations in the Microstructure of Cognition. MIT Press, MA.

[122]. Rumelhart, D.E., Widrow, B., Lehr, M.A., 1994. The basic ideas in neural networks. Communications of the ACM 37 (3), 87–92.

[123]. Rumelhart, D.E., Durbin, R., Golden, R., Chauvin, Y., 1995 Backpropagation: the basic theory. In: Chauvin, Y., Rumelhart D.E. (Eds.), Backpropagation: Theory, Architectures, and Applications. Lawrence Erlbaum Associates, New Jersey, pp 1–34.

[124]. Werbos, P.J., 1974. Beyond regression: new tools for prediction and analysis in the behavioral sciences. Ph.D. thesis, Harvard University

[125]. Werbos, P.J., 1988. Generalization of backpropagation with appli cation to a recurrent gas market model. Neural Networks 1, 339–356.

[126]. Lapedes, A., Farber, R., 1987. Nonlinear signal processing using neural networks: prediction and system modeling. Technical Report LA-UR-87-2662, Los Alamos National Laboratory, Los Alamos, NM.

[127]. Lapedes, A., Farber, R., 1988. How neural nets work. In Anderson, D.Z., (Ed.), Neural Information Processing Systems American Institute of Physics, New York, pp. 442–456.

[128]. Elsner, J. B., Tsonis, A., 1992, "Nonlinear Prediction Chaos and Noise", Bull. Amer. Meteor. Soc., 73, 49-60.

[129]. Weigend, A.S., Huberman, B.A., Rumelhart, D.E., 1990. Predicting the future: A connectionist approach. International Journal of Neural Systems 1, 193–209.

[130]. Weigend, A.S., Huberman, B.A., Rumelhart, D.E., 1992. Predicting sunspots and exchange rates with connectionist networks In: Casdagli, M., Eubank, S. (Eds.), Nonlinear Modeling and Forecasting. Addison-Wesley, Redwood City, CA, pp. 395–432.

[131]. Weigend, A.S., Rumelhart, D.E., Huberman, B.A., 1991. Generalization by weight-elimination with application to forecasting Advances in Neural Information Processing Systems 3, 875– 882.

[132]. Cottrell, M., Girard, B., Girard,Y., Mangeas, M., Muller, C., 1995. Neural modeling for time series: a statistical stepwise method for weight elimination. IEEE Transactions on Neural Networks 6 (6), 1355–1364.

[133]. Sharda, R., Patil, R.B., 1992. Connectionist approach to time series prediction: An empirical test. Journal of Intelligent Manufacturing 3, 317–323.

[134]. Tang, Z., Fishwick, P.A., 1993. Feedforward neural nets as models for time series forecasting. ORSA Journal on Computing 5 (4), 374–385.

[135]. Weigend, A.S., Gershenfeld, N.A., 1993. Time Series Prediction Forecasting the Future and Understanding the Past. Addison Wesley, Reading, MA

[136]. Karmakar, S., Kowar M,K,, Guhathakurta P., 2009, "Artificial Neural Network Skeleton in Deterministic "Forecast to Recognize Pattern of TMRF Ambikapur District", *CSVTU Research Journal, Bhilai, India*, 2 (2), 41-45.

[137]. Satish Kumar, "Neural Network Computer Engineering Series", The McGraw-Hill,New Delhi, 2007, 104 152.

[138]. Sivanandam, S.N., Sumathi, S., Deepa, S.N., "Introduction to Neural Network using MATLAB 6.0", *The McGraw-Hill,New Delhi*, 185-197.

[139]. Phillip H Sherrod, (2003), "DTREG Predictive Modeling Software", *www.dtreg.com*, 175-181.

[140]. Chatfield, C. and Collins, A.J., 1990, Introduction to multivariate Analysis, Chapman and Hall Publication.

[141]. Johnson, R.A. and Wichern, D.W., 1996, Applied multivariate statistical analysis, Prentice Hall of India private ltd.

[142]. Silva, F. de A. S. e. & Azevedo, C. A. V. de. Principal Components Analysis in the Software Assistat-Statistical Attendance. In:WORLD CONGRESS ON COMPUTERS IN

AGRICULTURE, 7, Reno-NV-USA: American Society of Agricultural and Biological Engineers, 2009.

[143]. Silva, F. de A. S. e. & Azevedo, C. A. V. de. A New Version of The Assistat-Statistical Assistance Software. In: World Congress On Computers In Agriculture, 4, Orlando-FL-USA: Anais... Orlando: American Society of Agricultural and Biological Engineers, 2006. pp.393-396.

[144]. Silva, F. de A. S. e. & Azevedo, C. A. V. de. Versão do programa computacional Assistat para o sistema operacional Windows. Revista Brasileira de Produtos Agroindustriais, Campina Grande, v.4,n.1, p71-78,2002.

[145]. Silva, F.de A.S.e. The ASSISTAT Software: statistical assistance. In: International Conference On Computers In Agriculture, 6, Cancun, 1996. Anais... Cancun: American Society of Agricultural Engineers, 1996. pp.294-298.

Abstract

Long-range monsoon rainfall data time series over a smaller geographical region like 'districts' is representing chaotic in nature. It is found that, identification of internal dynamics of its time series as well future prediction is very-very difficult. At present, it is a vital challenging task for meteorological services all over the world. It is found that, ANN technology has produced sufficient skill especially in prediction. Thus four separate, back-propagation neural network (BPN) models have been developed and verified. These models are BPN model in deterministic forecast, BPN model in parametric forecast, Principal Components BPN model and Hybrid BPN model. The model in deterministic forecast has produced excellent results, explained by strong relation between dependent (i.e., current year rainfall), and independent variables (i.e., past recorded rainfall data time series). Likewise, other models has also produced excellent results, explained by strong relation between their dependent (current year rainfall), and their corresponding independent variables (i.e., predictors). These models have been implemented through Java programming language. The entire development observations have been discussed in this book and expected that this observations will be extremely helpful for scientist and engineers those are actually tried to use BPN in their own similar applications.

Key Words: Neural Network, Back-Propagation, Chaotic Motion, Monsoon Rainfall Prediction, Meteorology.